Forest Business Management

AF663456

Forest Business Management

Dr. Girish B. Shahapurmath
M.Sc. (Forestry) Ph.D.
Assistant Professor of Forest Management
Department of Natural Resource Management
College of Forestry, Sirsi (UK), Karnataka

Dr. S. S. Inamati
M.Sc. (Forestry) Ph.D.
Assistant Professor of Silviculture and Agroforesry
Department of Silviculture and Agroforestry
College of Forestry, Sirsi (UK), Karnataka

SSPH

SATISH SERIAL PUBLISHING HOUSE
403, Express Tower, Commercial Complex
Azadpur, Delhi-110033 (India)
Phone : 011-27672852, Fax : 91-11-27672046
e-mail : info@satishserial.com, hkjain1975@yahoo.com
Website : www.satishserial.com

Published by :

SATISH SERIAL PUBLISHING HOUSE
403, Express Tower, Commercial Complex, Azadpur, Delhi-110033 (INDIA)
Phone : 011-27672852 Fax : 91-11-27672046
E-mail : info@satishserial.com, hkjain1975@yahoo.com

© **Publisher**

ISBN 978-93-88020-61-9

© 2020. All rights reserved, no part of this publication may be reproduced, stored in a retrieval system or transmitted in any form or by any means, electronic, mechanical, photocopying, recording or otherwise without the prior written permission of the publisher and also the copyright, rights of the printing, publishing, e-book of this edition and subsequent editions will vest with the publisher. All Computer floppies, CD's, e-book and in any other form relating to this book will be exclusive property of the publisher.

This book contains information obtained from authentic and highly regarded sources. Reasonable efforts have been made to publish reliable data and information, but the publisher cannot assume responsibility for the validity of all materials or the consequences of their use. The publisher have attempted to trace and acknowledge the copyright holders of all material reproduced in this publication and apologize to copyright holders if permission and acknowledgements to publish in this form have not been obtained. If any copyright material has not been acknowledged please write and let us know so that we may rectify it.

Composed, Designed & Printed in India

About the Authors

Dr. GIRISH B. SHAHAPURMATH, M.Sc. (Forestry) Ph.D.
Assistant Professor of Forest Management
Department of Natural Resource Management
College of Forestry, Sirsi-581 401, Karnataka
University of Agricultural Sciences, Dharwad, India
E-mail: girishbshahapur@gmail.com /
shahapurmathgb@uasd.in

- Specialization in Silviculture and Agroforestry, Forest Management, Remote Sensing and GIS.
- Post Graduate Diploma in Agricultural Extension and Management (PGDAEM) by MANAGAE, Hyderabad during 2014-15.
- Post Graduate Diploma Computer Applications (PGDCA) by Infotech Computer Education and Manipal Institute of Computer Education (MICE) during 2008-09.
- Total 16 years of experience in the field of forestry (Teaching, Research and Extension).
- Presently working as Assistant Professor of Forest Management in Department of Natural Resource Management at College of Forestry, SIRSI (University of Agricultural Sciences, Dharwad, Karnataka (India).
- Recipient of "Outstanding Scientist Award" in Forestry by The Society of Tropical Agriculture, New Delhi, India in 7th International Conference on Agriculture, Horticulture And Plant Sciences held at Shimla (H. P.) during June 28-29, 2018.
- Recipient of "Best Teacher Award-2016" for outstanding contribution in the field of Forestry on the occasion of International Conference on "Rural Livelihood Improvement by Enhancing Farmers' Income through Sustainable Innovative Agri and Allied Enterprises (RLSIAAe)" organized by the Society for Upliftment of Rural Economy, Varanasi (India) during 30th October to 1st November, 2018 held at Birla Institute of Technology, Patna.
- Recognized as a Life Member for the "Society for Upliftment of Rural Economy, Varanasi (Membership No. LM/245/2018).
- First prize for Best Poster Presentation on Multistoried Cropping of Teak-perennial vegetables-legume based Agroforestry system

Presented in National Symposium (Silver Jubilee) on "Agroforestry Knowledge for Sustainability, Climate Moderation and Challenges Ahead" held at National Research Centre for Agroforestry, Jhansi (UP) from 15-17 December, 2008.

- First Prize for Best Poster presentation on "Forest Law and Models" presented in "National Kaanoonu Utsava" Exhibition held at Sirsi from 07.06.2014 to 06.07.2014 organized by Karnataka High Court, Bengaluru and District Law Service, Karwar and Bar Association, Sirsi.
- Worked as Principal Investigator and CO Principal Investigator for 10 Research Projects.
- Field of interest - Silviculture and Agroforestry, Forest Management and Policy, Sustainable Forest Management, Remote Sensing and Geographical Information System in Forestry.
- Developed the course materials as reference books and for competitive exams, such as ASRB, IFS, ACF, RFO, KPSC, etc., in the field of forestry for students' purpose.
- Attended more than 15 Trainings/Summer/Winter schools, short courses, Seminars, Workshops, Conferences and undergone Agricultural Massive Open Online Courses (Ag MOOC).
- Published more 100 publications including Research papers in NAAS rated journal/Research notes/Papers presented and published in symposium/seminar/conference/workshop/Training/international conference/Popular articles published in journals/newspapers/Books/chapters in books.

Dr. S.S. INAMATI, M.Sc. (Forestry) Ph.D.
Assistant Professor of Silviculture and Agroforestry
Department of Silviculture and Agroforestry
College of Forestry, SIRSI-581 401, Karnataka
University of Agricultural Sciences, Dharwad, India
E-mail: inamatiss@gmail.com/inamatiss@uasd.in

- Post Graduate Diploma in Rural Development (PGDRD) by IGNOU, New Delhi during 2004-05.
- Post Graduate Diploma Environmental Sustainable Development (PGDESD) by IGNOU, New Delhi during 2018-19.
- Worked as Subject Matter Specialist (Agroforestry), KVK, Gnagavati from 21.01.2006 to 04.11.2008.
- Presently working as Assistant Professor of Silviculture and Agroforestry in the Deptt. of Silviculture and Agroforestry, College of Forestry, Sirsi Since 05.11.2008 till date (10 years)
- Total 15 years of experience in the field of forestry (Teaching, Research and Extension).
- Awarded "Best Presentation" in International Conference on Agroforestry and Forestry Technologies, Amsterdam, The Netherlands in during January 20-22, 2018.
- Research areas of Interest - Site specific Silviculture, Forest tree nursery management, Vegetational studies, Forest modelling.
- Developed e-course materials and practical manuals for various courses of Silviculture and Agroforestry, Lecture Notes on Forest Mensuration, Lecture Notes Silviculture of Indian Trees, Lecture Notes on Quantitative Silviculture, Lecture Notes Productivity of Agroforestry systems.
- Attended more than 35 Conferences, Workshops, Trainings, Winter schools and short courses.
- Worked as Principal Investigator and CO-Principal Investigator for 12 Research Projects.
- Published more than 140 publications including Research papers in NAAS rated journal/Research notes/Papers presented and published in symposium/seminar/conference/workshop/Training/international conference/Popular articles published in journals/newspapers/Books/chapters in books.

Hon'ble
Vice Chancellor

UNIVERSITY OF AGRICULTURAL SCIENCES, DHARWAD
KRISHI NAGAR, DHARWAD-581 005
KARNATAKA STATE, INDIA

Foreword

The term 'business' should be used to convey the same meaning as the term 'trade' simply denotes purchase and sale of goods, whereas 'business' includes all activities form production to distribution of goods and services. It embraces industry, trade and other activities like banking, transport, Insurance and warehousing which facilitates production and distribution of goods and services. According to F.C. Hopper "The whole complex field of commerce and industry which includes the basic industries, processing and manufacturing industries, and the network of ancillary services: distribution, banking, insurance transport and so on, which serve and inter penetrate the world of business as a whole" are called business activities.

Forestry in India is a significant rural industry and a major environmental resource. India is one of the ten most forest-rich countries of the world along with the Russian Federation, Brazil, Canada, United States of America, China, Democratic Republic of the Congo, Australia, Indonesia and Sudan. Together, India and these countries account for 67 per cent of total forest area of the world. India's forest cover grew at 0.22% annually over 1990-2000, and has grown at the rate of 0.46% per year over 2000-2010, after decades where forest degradation was a matter of serious concern.

As of 2010, the Food and Agriculture Organisation of the United Nations estimates India's forest cover to be about 68 million hectares, or 22% of the country's area. The 2013 Forest Survey of India states its forest cover increased to 69.8 million hectares by 2012, per satellite measurements; this represents an increase of 5,871 square kilometres of forest cover in 2 years. However, the gains were primarily in northern, central and southern Indian states, while northeastern states witnessed a net loss in forest cover over 2010 to 2012.

In 2002, forestry industry contributed 1.7% to India's GDP. In 2010, the contribution to GDP dropped to 0.9%, largely because of rapid growth of the economy in other sectors and the government's decision to reform and reduce import tariffs to let imports satisfy the growing Indian demand for wood products. India produces a range of processed forest (wood and non-wood) products ranging from wood panel products and wood pulp. India's paper industry produces over 3,000 metric tonnes annually from more than 400 mills. The furniture and craft industry is another consumer of wood. India's wood-based processing industries consumed about 30 million cubic metres of industrial wood in 2002. India annually consumes an additional 270 million tonnes of fuelwood, 2800 million tonnes of fodder, and about 102 million cubic metre of forest products - valued at about ₹ 27,500 crore (US$4.1 billion) a year.

India is one of the world's largest consumer of fuel-wood. India's consumption of fuel-wood is about five times higher than what can be sustainably removed from forests. However, a large percentage of this fuel-wood is grown as biomass remaining from agriculture, and is managed outside forests. Fuel-wood meets about 40% of the energy needs of the country. Around 80% of rural people and 48% of urban people use fuel-wood. Unless India makes major, rapid and sustained effort to expand electricity generation and power plants, the rural and urban poor in India will continue to meet their energy needs through unsustainable destruction of forests and fuel-wood consumption. India's dependence on fuel-wood and forestry products as a primary energy source is not only environmentally unsustainable, it is a primary cause of India's near-permanent haze and air pollution.

Forestry in India is more than just about wood and fuel. India has a thriving non-wood forest products industry, which produces latex, gums, resins, essential oils, flavours, fragrances and aroma chemicals, incense sticks, handicrafts, thatching materials and medicinal plants. About 60% of non-wood forest products production is consumed locally. About 50% of the total revenue from the forestry industry in India is in non-wood forest products category. In 2002, non-wood forest products were a source of significant supplemental income to over 400 million people in India, mostly rural.

Textbooks on Business Management in Forestry suitable for undergraduate and post graduate students are very few particularly in Forestry Colleges in India. This textbook on "Forest Business Management" by Dr. Girish B. Shahapurmath and Dr. S. S. Inamati is a timely and welcome contribution in the forestry field. The authors have a considerable experience in teaching, research and extension in forestry/agroforestry and have presented the subject matter critically and systematically. This book will serve as a valuable source of information for the academicians in forestry institutes and also for the students for the preparation of competitive examinations in forestry discipline. I congratulate the authors for the hard work they put in, the systematic selection of topics and well documented and meticulous presentation.

I am hopeful that this text book will adequately meet the needs of students of forestry colleges and would be widely welcomed. I extend my best wishes to Dr. Girish B. Shahapurmath and Dr. S. S. Inamati for the success of their book.

May, 2019
Dharwad, Karnataka

Dr. M.B. CHETTI
(Vice Chancellor)
University of Agricultural Sciences, Dharwad
and Former Assistant Director General (HRD)
ICAR, New Delhi

Preface

Forestry, as a subject, is being taught to students of the Forest Colleges in different Agricultural Universities in India. This subject had been introduced as a subject for degree courses in most of the Indian Agricultural Universities. Indian Council of Agricultural Research (ICAR) provided an opportunity to start state-run forestry colleges in various State Agricultural Universities (SAUs) in the country. Earlier as per the recommendation of the National Commission on Agriculture, ICAR had started a course in 'Farm Forestry' during 1976 in all SAUs. There was a growing concern among the academicians and politicians alike that the students of forestry should have a greater exposure and access to natural forests. There has been increased public awareness of natural resource policy, including forest management. Forestry in India is more than just about wood and fuel. India has a thriving non-wood forest products industry, which produces latex, gums, resins, essential oils, flavours, fragrances and aroma chemicals, incense sticks, handicrafts, thatching materials and medicinal plants. About 60% of non-wood forest products production is consumed locally. About 50% of the total revenue from the forestry industry in India is in non-wood forest products category. In 2002, non-wood forest products were a source of significant supplemental income to over 400 million people in India, mostly rural. Therefore, this book has been written to meet the requirements of the students of the Forest Colleges as well as those of the students preparing for competitive examinations.

The book covers mostly the present syllabi of the Forest Colleges as well as those of the competitive examinations (Indian Forest Service and State Forest Services). I claim no originality in writing this book except for the arrangement and presentation of the subject matter. A large number of books, journals and proceedings of symposia and workshops have been consulted. A list of these references is given at the end of book as references / bibliography.

I, therefore, gratefully acknowledge the help taken from "Fundamentals of Farm Business Management by S.S. Johl and T.R. Kapur and other publications mentioned in the references.

I am greatly indebted to Dean (Forestry) for constant guidance and help during the time of writing the book and improving it substantially. I am equally grateful to all the faculty members of College of Forestry, Sirsi (UAS, Dharwad, Karnataka) and officers of Karnataka Forest

Department for their valuable suggestions and help in writing the book as well as for reading through the chapters and making very valuable improvements.

Place: Sirsi, Karnataka
Date: 1st January, 2019

Dr. Girish B. Shahapurmath
Assistant Professor of Forest Management
Department of Natural Resource Managment
College of Forestry, Sirsi (UAS, Dharwad)
Karnataka, India

Contents

Chapter 1

Forest Business Management - Definition, Scope, Importance and Characteristics of Forest Business Management

Farm Management

Farm management comprises of two words "**farm**" and "**management**". Literally "**farm**" means a piece of land where crops and livestock enterprises are taken up under a common management and has specific boundaries. "**Management**" means the act or art of managing. The art of managing a farm successfully, as measured by the test of profitableness is called farm management.

In simple words, the daily work of supervision of farm labour and carrying out the directives of seniors by the public or private employed farm manager is generally referred to as farm management.

Definition of Farm Management

Management is a widely used term but which is subject to many individual definitions. Of the many Farm Management definitions, if we take one:

"Farm Management is the decision-making process whereby limited resources are allocated to a number or production alternatives to organise and operate the business in such a way as to attain some objectives." Some of the definitions given by eminent farm Management authorities are: -

1. Farm Management is that branch of agricultural economics, which deals with the business principles, and practices of farming with an object of obtaining the maximum possible return from the farm as a unit under a sound-farming programme.

2. "Farm Management" is the study of the business principle in farming. It may be defined as the science of organisation and the management for continuous profit" - Warren
3. "It is the science which considers the organization and operation of the farm from the point of view of efficiency and continuous profit"- Effersen.
4. "Farm Management as the sub-division of economics which considers the allocation of limited resources within the individual farm is a science of choice and decision making, and thus is a field requiring studied judgement" Heady and Jencon.

   Farm management has become complex due to :

- Frequent price changes in agri-commodities
- New Technology
  - Development of new seed varieties, new fertiliser.
  - New breeds of cattle
  - New chemicals for pests and weed control
  - New animal health product and feed activities.

Changes like machinery, irrigation equipment, computers, Govt. Policies. These are some of changing environment affecting the manager, and as a result the production is affected. Some managers respond it correctly, some incorrectly to the technology, price and other economic factors.

Role of Farm Management

The main role of farm management is to help in realisation of the maximum not profit from the various enterprises on a farm. A typical farm is a combination of two or more enterprises and the chief aim of the farm manager would be to get the whole unit give the maximum total returns. It is not the return from anyone enterprise that determine the financial success of a farm, but it is the total return from all enterprises that counts its success or failure. A proper understanding of farm management principle helps in selection combination and execution of enterprises, which are consistent to a sound agricultural policy. The management study is undertaken:

1. To study the input-output relationship in agriculture and determine the relative efficiency of various factor combination.

2. To determine the most the profitable crop production and livestock raising methods.
3. To study the cost per hectare and per quintal or per kg.
4. To evaluate the farm resource and land use.
5. To study the comparative economics of different enterprises.
6. To determine the relation of size of farm to land utilisation, cropping pattern, capital investment and labour employment.
7. To study the impact of technical changes on farm business.
8. To find out ways and means for increasing the efficiency of arm business through better input-output relationship and proper allocation of resources among different uses.

Forest Management is defined as the practical application of the scientific, technical and economic principles of forestry. Forest management is that branch of forestry whose function is the organization of a forest property for management and maintenance by ordering in time and place the various operations necessary for the conservation, protection and improvement of the forest on the one hand and the controlled harvesting of the forest on the other.

Management of forests broadly involves three main tasks:

1. Control of composition and structure of the growing stock
2. Harvesting and marketing of forest produce
3. Administration of forest property and personnel

An illustration of Farm Management Process (Diagram)

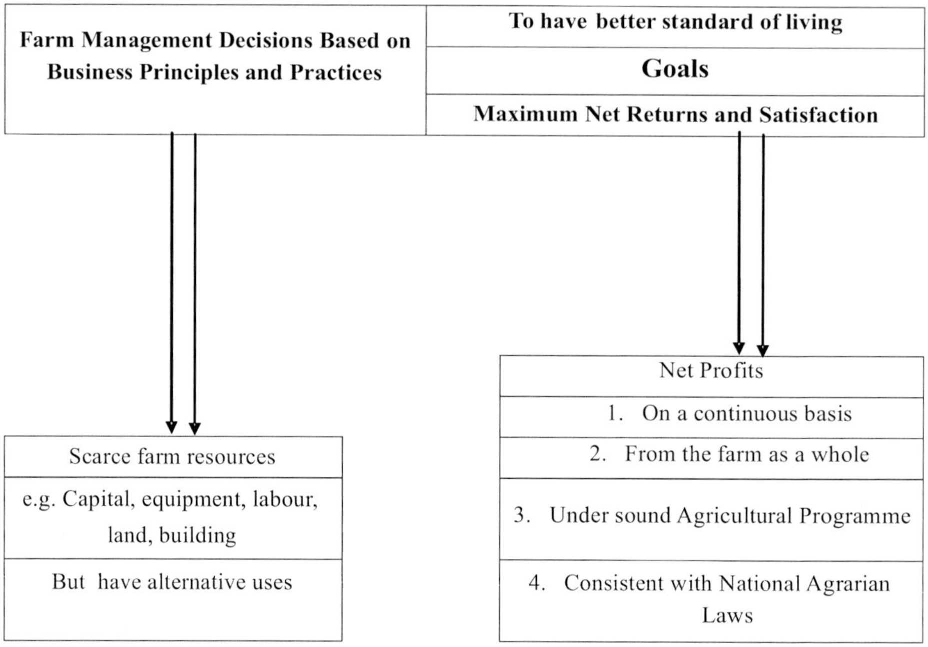

In simple words, farm management can be defined as a science which deals with judicious decisions on the use of scarce farm resources, having alternative uses to obtain the maximum profit and family satisfaction on a continuous basis from the farm as a whole and under sound farming programmes. In other words, farm management seeks to help the farmer in deciding problems like what to produce, how much to produce, how to produce and when to buy and sell and in organization and managerial problems relating to these decisions. The goal of increasing income and leisure also depends upon the "**planning horizon**" and personal values of the farmer. Planning horizon **is defined as the period covered by a particular plan or a firm's planning cycle. In general, its length is dictated by the degree of uncertainty in the external environment: higher the uncertainty, shorter the planning horizon.**

Farm management may in short be called as a science of decision-making or a science of choice as a continuous process. The principal changes frequently encountered in the farm or outside the farm by the farmer are: fluctuations in prices; weather variations; inventions in farming methods; changes in socio-economic environment including changes in Govt. policy and social responses and values.

What makes a successful Farm Manager?

Successful farm management requires the ability and capacity to make the correct decisions. **Nielson proposed eight processes or functions to improve the management ability**; one needs a process model to know what one is trying to improve.

1. Formulation of the goals or objectives of the firm
2. Recognition and definition of a problem or opportunity
3. Obtaining information-observation of relevant facts
4. Specification and analysis of alternatives
5. Decision-making choosing an alternative
6. Taking action
7. Bearing responsibility for the decision or action taken
8. Evaluating the outcome

Scope of Farm Management

Farm management is generally considered to fall in the field of microeconomics. It deals with the allocation of resources at the level of an individual farm and the primary concern of farm management is the farm as a unit. **The subject of farm management includes: Farm management research, training and extension.**

Farm Management Research : Solution of economic problems faced by a farmer is greatly facilitated through recording of data relating to the farm. Remedial measures have to be formulated after analyzing these data. Although the problems and their solutions will vary from time to time, in the developing countries the following aspects need to be continuously researched:

a) Delineation of homogeneous type of farming areas in various regions of the country.
b) Generation of input-output coefficients and working out comparative economics of various farm enterprises.
c) Formulation of standard farm plans and optimum cropping patterns for different areas and types of farming.
d) Developing suitable models of mechanization and modernization.
e) Evaluation of agricultural policies having a bearing on development and growth of the farm firms.

Farm Management Teaching : Almost all Agricultural Universities in India have one or more compulsory courses in farm management for undergraduate students. It is being realized that the training in farm management science is essential for agricultural graduates for understanding farmers' response to varying economic pressures and stimuli. With the new technological break through in agriculture, the decision-making processes have become more complex on farm organizations. To help the farmer take right decisions as to what to grow, how much to grow, how to grow, when and where to buy and sell, higher training in this subject has become necessary. With this background Indian Agricultural Research Institute (IARI), New Delhi and many Agricultural Universities in India have started teaching advanced farm management as major subject at M.Sc. and Ph.D. level.

Farm Management Extension : The managerial ability of farmers can be improved appreciably through extension education programmes in farm management. Short course in farm management for progressive farmers engaged in various types of farm enterprises and for young farmers can be organized. Extension specialists in farm management can improve the managerial ability of the farmers through district, block and village level training camps. Research, teaching and extension together, thus, seek improving the ability of the farmers to introduce desirable changes in the utilization of scarce resources at the farm with a view to increase incomes and improve standards of living of the farmers.

Importance of the subject of farm management in India

a) Scope of increase in farm income through suitable farm adjustments

b) Technological progress and role or farm management

c) Industrial development and farm management

Farm management as an educational tool

Level of awareness of farmers increases with increased education facilities and extension programmes. To make full use of rapidly increasing new technology, the adoption process of agricultural processes need to accelerate. Farmers learn more quickly, if they are convinced on cost returns analysis of each suggested management action. Economic analysis makes the farmers learn more of alternative courses of action. The scientific management process acts as a useful educational tool through gathering more information on new alternatives and testing each recommendation on economic standards.

Agribusiness

John H. Davis of Harvard University first used the term 'agribusiness' in 1955. In 1980s, it was given three connotations: (1) synonymous with term agriculture, (2) synonymous with agricultural economics and (3) a modified concept of agriculture, excluding farming, or the off-farm aspects of agriculture.

At present, agribusiness is defined as all business enterprises or sells to farmers/traders/consumers. **The transaction may involve either an input or a produce or service and encompasses items, such as:**

2 Productive resources (feed, seed, fertilizer, equipment, energy, pesticides, machinery, etc.)

3 Agricultural commodities - (raw and processed commodities of food and fibre)

4 Facilitative services (credit, insurance, marketing, storage, processing, transportation, packing, distribution, consultancy, soil testing, etc.).

Scope for Agribusiness in India (Agri, Horti, Forestry, Sericulture, Apiculture, Mushroom, Organic farming, Animal husbandry, etc.)

1. India is endowed with varied agro-climate, which facilitates production of temperate, sub-tropical and tropical agricultural commodities.
2. There is growing demand for agricultural inputs like seed, fodder, inorganic fertilizers and bio-fertilizers.
3. Biotechnology applications in agriculture have vast scope in production of seed, bio-control agents, industrial harnessing of microbes for bakery products.
4. Export can be harnessed as a source of economic growth. As a signatory of World Trade Organization, India has vast potential to improve its present position in the World trade of agricultural

commodities both raw and processed form. The products line include cereals, pulses, oilseeds and oils, oil meal, spices and condiments, fruits and vegetables, flowers, medicinal plants and essential oils, agricultural advisory services, agricultural tools and implements, meat, milk and milk products, fish and fish products, ornamental fish, forest by products, etc.

5. At present processing is done at primary level only and the rising standard of living expands opportunities for secondary and tertiary processing of agricultural commodities.
6. The vast coastal line and internal water courses provides enormous opportunity for production of marine and inland fish and ornamental fish culture gaining popularity with increase in aesthetic value among the citizens of India.
7. The livestock wealth gives enormous scope for production of meat, milk and milk products, poultry products, etc.
8. The forest resources can be utilized for production of by-products of forestry.
9. Beekeeping and apiary can be taken up on large scale in India.
10. Mushroom production for domestic consumption and export can be enhanced with improvement in the state of art of their production.
11. Organic farming has highest potential in India as the pesticide and inorganic fertilizer application are less in India compared to industrial nations of the world. The farmers can be encouraged and educated to switch over for organic farming.
12. There is wide scope for production and promotion of bio-pesticides and bio-control agents for protection of crops.
13. Seeds, hybrid and genetically modified crops, have the highest potential in India in the future, since the productivity of high yielding varieties have reached a plateau.
14. Micro-irrigation systems and labor saving farm equipments have good potential for the years to come due to declining groundwater level and labor scarcity for agricultural operations like weeding, transplanting and harvesting.
15. Production of vegetables and flowers under green house conditions can be taken up to harness the export market.

16. Trained human resources in agriculture and allied sciences will take on agricultural extension system due to dwindling resources of state finance and downsizing the present government agricultural extension staff as consulting services.
17. The enhanced agricultural production throws open opportunities for employment in marketing, transport, cold storage and warehousing facilities, credit, insurance and logistic support services.

Types of Small Businesses

With the exception of Government, most of the small businesses can be classified as the following types :

1. Production
2. Retailing
3. Distribution
4. Personal services
5. Professional services
6. Financial
7. Franchising

1. **Production :** This classification includes all types of production including agricultural production of crops and livestock, as well as forestry.
2. **Distribution :** This classification refers to those businesses, which do not make anything but which bring the goods and services to the consumer or user. This includes such activities as packaging, labelling, transporting, refrigerating, freezing, processing, storing, and performing any service necessary to prepare the goods or to provide the service to eventual consumer.
3. **Retailing :** Although often included as a **phase of distribution**, retailing is listed as a separate category because there are a large number of persons employed in retailing. Obviously, it represents one of the best opportunities for the potential entrepreneur. Retailing is that **stage of distribution**, which deals with the consumers. Examples of retailers are grocers, self-service stores, florists, agricultural input retailing.

4. **Personal services :** The personal service business are those, which do not primarily supply goods to the public, but instead perform a service. Goods may be used to perform the service but they are of secondary importance. Examples of personal service are hotels, restaurants, agro-service centers.
5. **Professional services :** Some type of services, in order to protect the public, requires considerable training on the part of those offering the service. Usually that professional services must have a formal education and rigid examinations before receiving licenses to offer their services to the public. Examples of those offering services are investment brokers, insurance agents, etc.
6. **Financial :** Financial businesses are usually service-oriented but since they deal primarily with the loaning or investing of money or the equivalents of money (stocks, bonds, property rights, etc.) a separate category describes them best. Examples of financial services are commercial banks, insurance companies, thrift and loan societies, etc.
7. **Franchising :** Franchising is a system for selectively distributing goods or services through outlets owned by the franchisee. Basically, a franchise is a patent or trademark, license, entitling the holder to market particular products or services under a brand name or trademark according to pre-arranged terms and conditions. The franchiser is the owner of his or her own business (the franchisee) is likely to be more diligent and strive harder for success than the hired manager of a company-owned outlet. Since franchising is form of selective distribution, the typical franchise agreement prohibits the franchise from setting up competing outlets within the franchise area. Examples of franchise services are diet services, quick-service food-drive inns like fried chicken.

Nature and characteristics of farm management science

The farm management science has many distinguishing characteristics from other fields of agricultural science. A few important characteristics are as follows:

1. Practical Science
2. Profit Oriented
3. Integrating Science

4. Broader Field
5. Micro-approach
6. Farm Unit as Whole

1. **Practical Science :** Farm management is a practical science which aims at testing the applicability of facts of other physical and biological sciences and findings and showing how to put these results to use on a given farm situation. A farmer has to select a method which is more practicable and economical to his particular farm situation taking into consideration the volume of work and financial implications. The acceptability of the facts of other physical and biological sciences are tested on the farm and determine whether those are economical and practicable on a given farm situation. Thus, farm management is a practical science.
2. **Profit Oriented :** The main objective of farm management is to earn maximum profit and hence, this science aims to have maximum economic efficiency rather than physical efficiency. Thus, farm management science is profit oriented. Farming is the livelihood of farmers. The net income generated from the farm should be able to meet the consumption, savings and investment needs of the farm family. Hence, the major objective of the farm management is to maximize profits from various activities in the farm.
3. **Integrating Science :** While operating the farm, number of findings of other sciences is actually used. Thus, farm manager has to co-ordinate all the findings of other sciences.

Integration of enterprises in the farm : Crop cultivation, milch cows, goat, sheep, poultry, and silkworm rearing, bee keeping, etc., are some of the enterprises commonly observed in the farms. These enterprises have to be integrated by using the output/residues from one enterprise as input for the other (e.g. fodder for cattle and cow dung as manure). This integration has to happen in a scientific manner looking at the farm in a holistic manner, with focus on economic efficiency.

4. **Broader Field :** The farm management specialist required to have detail information from other Sciences. Hence, for successful farming information of one or two sciences is not sufficient, as this science is too much broader field.

5. **Micro-approach :** Since this science is related to individual farm, it treats every farm unit unique in available resources. Each farm unit, therefore, has to be studied and planned individually.

6. **Farm Unit as Whole :** In farm management analysis a farm as a whole is considered to be the unit for making decisions because the objective is to maximize the returns from the whole farm instead of only improving the returns from a particular enterprise or a practice. Farm manager has to be a generalist who should be aware of technical aspects of crop cultivation and a business manager with skills in negotiation, marketing, human resource management, financial management, etc.

Relationship of Farm Management with Other Sciences

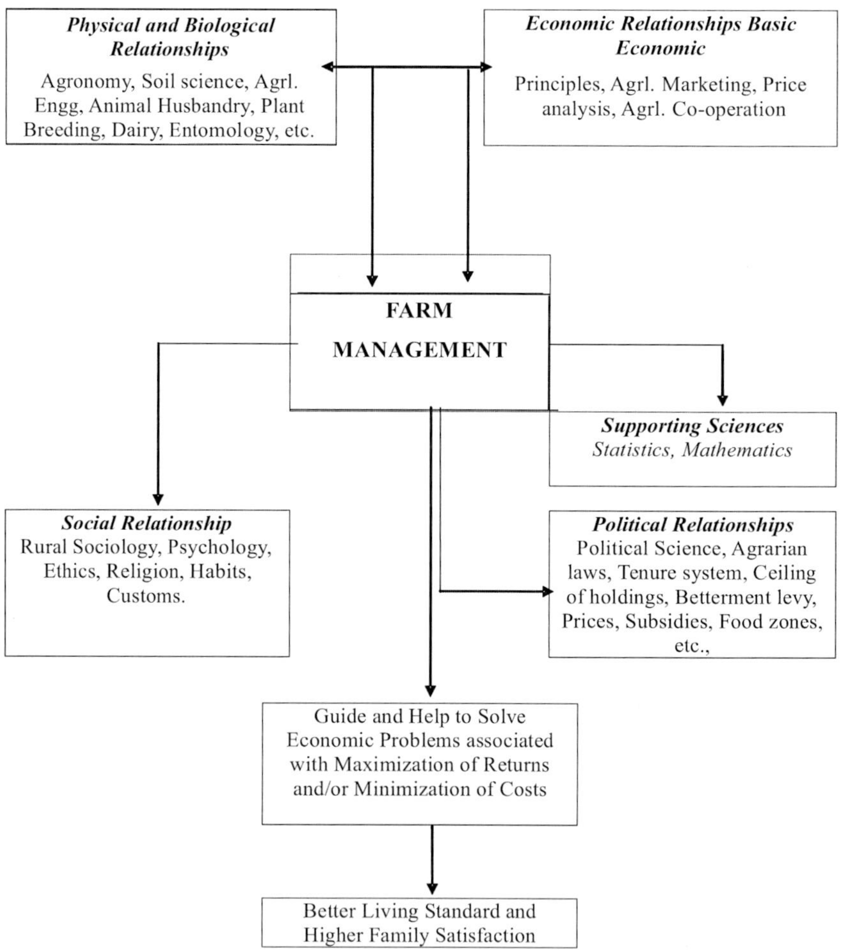

Farm Management Decisions Chart

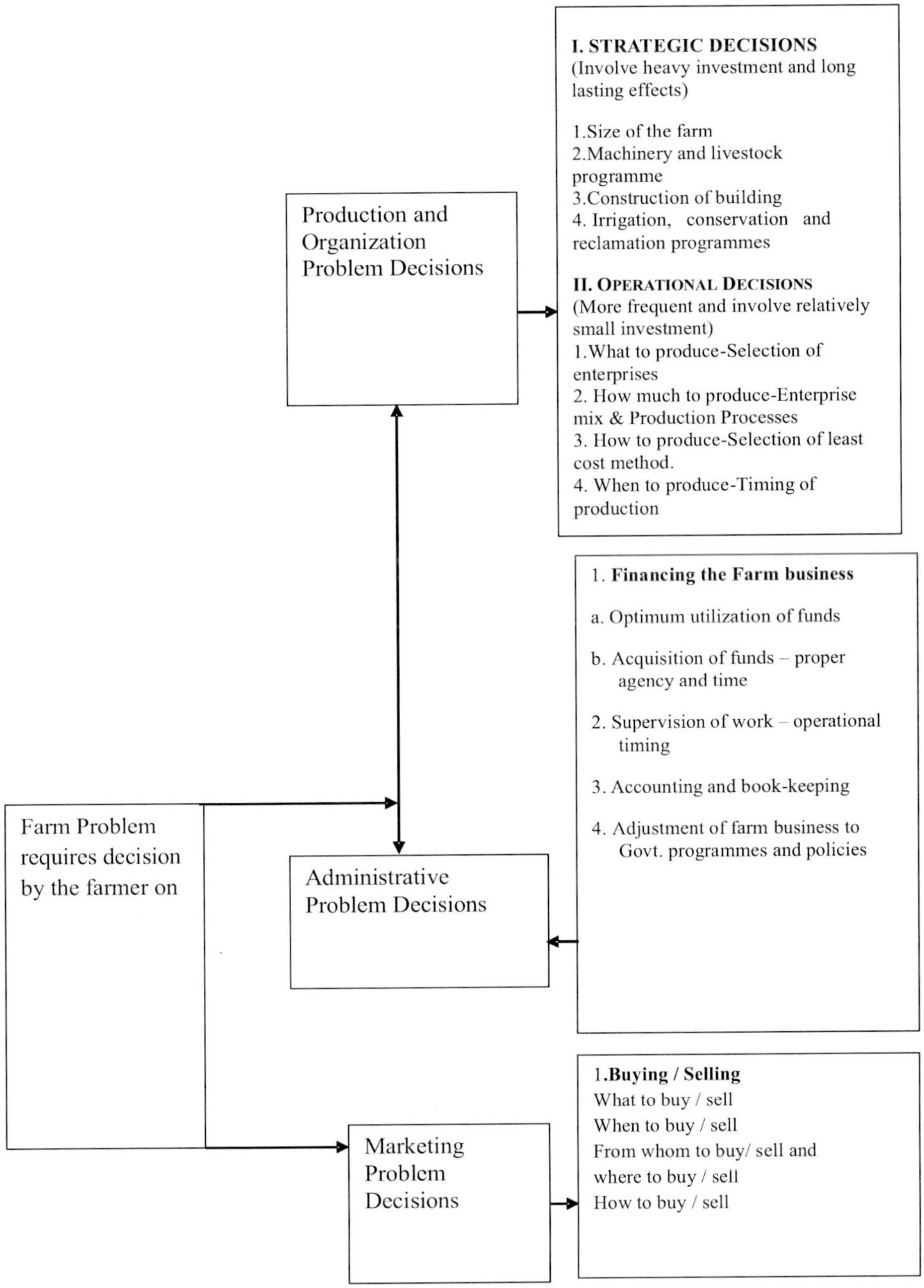

Fig.2.1 Farm-Business Decision-Making

The Scope of the Manager or Farm Management Activities

Once the management decision is made, it has to be translated into activities, or most likely, a series of activities. In agricultural and forestry business management, such farm management activities may be grouped into technical, commercial, financial, and accounting categories :

a) **Technical Activities :** These are comprised of all physical production activities. That is include responsibility for all production know how, seeing that productions accomplished on time, and adapting production processes to changing economic and technical conditions.

b) **Commercial Activities :** These include all buying and selling activities of the farm/business. This are involves procurement of inputs in the quantities and combinations necessary for efficient production, plus orderly storage, handling, marketing of commodities produced. It also includes the task of market forecasting and contracting for services of others.

c) **Financial Activities :** These are comprised of all activities related to the procurement and use of financial or money capital. That is acquisition and use of capital, presumably in an optimal manner. This requires forecasting future investment needs and arranging for their financing. Such capital is normally used for the purchase of physical assets and running the farm/business.

d) **Accounting :** These activities involve the establishment and maintenance of farm/business records, including records of buying and selling, etc. Include physical, human, business, and tax records. This area may involve setting standards for certain enterprises or segments of the business.

Table 1. Classification of Agricultural and Forestry Business Activities.

Classification of Activities	Description/Nature of Activities	Examples of Actual Decisions Made and Implemented.
Technical	Production of goods and services Using land Determining level of mechanization	What to produce: enterprise choice and combination? How much to produce: Amount to produce of each crop? How to produce: Amounts of inputs to use on each crop? How to combine them, how to co-ordinate their production? Capability-fertility Tillage practices- conservation Regulations-constraints Capital requirements Availability of services Labour implications
Commercial	Acquiring inputs Marketing products Forecasting costs and prices.	Purchase of inputs (source, terms, quality, quanity, financing, etc.) Sale of produce (form of produce; time to sell; place to sell; type of market- open market, contract, etc. Inputs and Outputs
Financial	Acquiring funds Using funds Forecasting future needs	Purpose of funds/ quantity and terms Source of financing Lender position Equity position and liquidity position Repayment plan Depreciation of assets Expansion / contraction Changing technology
Accounting	Keeping production records Recording commercial and financial transactions Tax reporting	Records of output of produce Records of use of resources Records of sales and purchases Records of borrowing and repayments Cash flow forecasting Income and other taxes, wages and depreciation

Characteristics of forest business management

1. Primary force of production
2. Size of the production unit:
3. Heavy dependence on climatic factors
4. Frequency and speed of decisions
5. Changes in prices
6. Standardization of practices and products
7. Turn over
8. Financing
9. Inseparable home and farm business
10. Nature of variable and fixed costs
11. In-elastic demand for forest products and
12. Time rigidity in consumption of forest products

1. **Primary force of production :** The farmers have a little control over production due to agriculture being seasonal and biological in character. Therefore, management practices in farming must be adjusted to meet these conditions. Agriculture requires a far larger proportion of land in relation to other factors of production than industry. This is the underlying cause of many of the chief differences between agriculture and industry, such as: **(i) diminishing returns at an earlier stage, (ii) wide scatter of production and (iii) greater importance of systems of land tenure.**

2. **Size of the production unit :** Farming is comparatively a small sized business and thus, provides little scope for division of labour. The farmer acts like both as a labourer and as a capitalist on his own farm on which all his accumulated wealth and annual savings get invested. However, most of other industries are organized on a large scale. Because of this difference in size of the unit, the management problems of agriculture vary greatly from other industries.

   Farming is not basically suited to large scale operations on a scale obtaining in industries, because it is concerned with living things, plants and animals scattered over space, which need prompt attention and personal care of the farmer.

3. **Heavy dependence on climatic factors :** Weather is a very important and unpredictable variable in all farming operations. Any sudden change in the weather temperature, rainfall or humidity leads to re-scheduling of the activities fixed for that day or week on farm. But in most of other business or industries, weather has a least importance.

   Farming is a business of great risk and uncertainty on account of excess or failure of rainfall, drought, hailstorm, floods insect pests and diseases. Therefore, management practice in farming must be more flexible than in other industries.

4. **Frequency and speed of decisions :** Farming requires many and speedy decisions on the part of the farmer and farm workers. When a sudden rain floods all the crop land, the fields must be drained immediately to save the growing crops, even if it takes all the night to do the job and costs tremendously.

5. **Changes in prices :** Agricultural prices and production usually move in opposite direction. Short run agriculture has a little control over the volume of production resulting in inelasticity of supplies and a smaller volume results in increase in prices. Demand for agricultural or forest products is also relatively more inelastic. Most other industries can apply checks over production by adjusting their manufacturing schedules. It is very difficult to adjust production to price variation in long run also.
6. **Standardization of practices and products :** Most of the non agricultural industries are characterized by the production of a large volume of highly standardized single line products by the use of machines and trained personnel. But such standardization is not possible in agriculture or forestry due to the difficulty in production and grading of absolutely uniform products. The management problems of the farmers differ from other industries because of the marketing difficulties involved in small volumes of indifferent quality products.
7. **Turn over :** Rate of capital turnover is the gross returns on a farm as a percentage of total capital investment:

$$\frac{\text{Gross income}}{\text{Total farm Assets}} \times 100 = \textbf{Rate of capital turn over}$$

Rate of capital turnover is the basic measure of efficiency in the use of capital employed by the business. This means the speed with which the original capital investment is re-captured by receipts from the business. A higher turnover rate usually means efficient use of capital. The turnover rate in agriculture or forestry is relatively slow because the production process ordinarily requires a period of several years. The management problems of farmers in regard to capital use differ greatly from industries. That is why the credit and financing problems of agriculture or forestry must be approached in a different manner.

8. **Financing :** Investment in agriculture is more risky and expensive as farming is subjected to many risks from storms, dry periods, insect pests, diseases and other calamities. So, agricultural financing faces different problems as compared to other types of business. And also repayment provisions are different due to slower rate of turnover. The recovery rate on the capital used in agriculture which is generally periodic and spreads over a longer period.

9. **Inseparable home and farm business :** In agriculture farm inputs may be partly used in the home, such as food grams, feed grains, transport vehicles, etc. Therefore, agriculture often regarded as a way of life as well as a means of livelihood such that socio-psychological and financial considerations influence its organization more than they do in other forms of industries and business.

10. **Nature of variable and fixed costs :** Of the total costs, portion of fixed costs is more in agriculture than in industry. This high proportion of fixed costs tends to make adjustments in production is more difficult. There, thus, enter rigidities in the production process and products. Their level and combinations cannot be changed quickly in response to changes in prices and economic environment.

11. **In-elastic demand for forest products :** As the standard of living improves and incomes increase, the demand for agricultural products will increase less rapidly than that for industrial products. In the lower level, demand for agricultural products (especially food) is inelastic, but in the upper portion the curve becomes more elastic. This kink comes at a point where reasonable level of consumption (C_R) is reached.

    This characteristic puts agriculture at a disadvantage. Higher production may reduce prices so low that total returns might not increase or may even decrease. On the other hand, if increased production comes, the decreasing marginal returns phase of the response curve, costs will go high.

12. **Time rigidity in consumption of forest products :** Utilization of agricultural products cannot be postponed, as they are generally perishable. Because of the small scale individual productions, the market intermediaries between the producers and the final consumers assume a place of particular importance. They come into provide place, time and form utility to the farm produce to make it available to the consumer in the form, at the time and place required by them. Therefore, the share of the producer in consumer's rupee goes low in case of farm forestry products.

❑❑❑

Chapter 2

Basics of Forest Production Decisions: Objectives of Production Economics

Forest / Farm Production Economics: Basic Concepts

1. **Production**: The process through which some goods and services called inputs are transformed into other goods called products or output.
2. **Production function**: A systematic and mathematical expression of the relationship among various quantities of inputs or input services used in the production of a commodity and the corresponding quantities of output is called a production function.
3. **Continuous production function**: This function arises for those inputs which can be divided into smaller doses. Continuous variables can be known from measurement, for example, seeds and fertilizers, etc.
4. **Discontinuous or discrete production function**: This function arises for those inputs or work units which cannot be divided into smaller units and hence, are used in whole numbers. For example, number of ploughings, weedings and harvestings, etc.
5. **Short run production period**: The planning period during which one or more of the resources are fixed while others are variable resources. The output can be varied only by intensive use of fixed resources. It is written as Y=f (X1, X2 / X3.....Xn), where Y is output, X1, X2 are variable inputs and X3.....Xn are fixed inputs.
6. **Long run production period**: The planning period during which all the resources can be varied. It is written as Y=f (X1, X2 ,.....Xn)
7. **Technical coefficient**: The amount of input per unit of output is called technical coefficient.

8. **Resources**: Anything that aids in production is called a resource. The resources physically enter the production process.
9. **Resource services**: The work done by a person, machine or livestock is called a resource service. Resources do not enter the production process physically.
10. **Fixed resources**: The resources that remain unchanged irrespective of the level of production are called fixed resources. For example land, building, machinery. These resources exist only in short run. The costs associated with these resources are called fixed costs.
11. **Variable resources**: The resources that vary with the level of production are called variable resources. These resources exist both in short run and long run. For example seeds, fertilizers, chemicals, etc. The costs associated with these resources are called variable costs.
12. **Flow resources**: The resources that cannot be stored and should be used as and when these are available. For example, services of a labourer on a particular day.
13. **Stock resources**: The resources that can be stored for use later on. For example seeds. Defining an input as a flow or stock depends on the length of time under consideration. For example, tractor with 10 years life is a stock resources, if we take the services of tractor for its entire useful life of 10 years. But it also provides its service every day, therefore it is a flow resources.
14. **Production period**: It is the time period required for the transformation of resources or inputs into products.
15. **Farm entrepreneur**: Farm entrepreneur is the person who organizes and operates the farm business and bears the responsibility of the outcome of the business.
16. **Farm business manager**: Person appointed by the entrepreneur to manage and supervise the farm business and is paid for the services rendered. He/she carries out the instructions of the entrepreneur.
17. **Productivity**: Output per unit of inputs is called the productivity.
18. **Technical efficiency**: It is the ratio of the physical output to inputs used. It implies the using of resources as effectively as possible without any wastages.

19. **Economic efficiency**: It is the expression of technical efficiency in monetary terms through the prices. In other words, the ratio of value of output to value of inputs is termed as economic efficiency. It implies maximization of profits per unit of input.

20. **Allocative efficiency**: It occurs when no possible re-organization of resources/production can make any combination higher yielding without making other combination less yielding. It refers to resource use efficiency.

21. **Optimality**: It is an ideal condition or situation in which costs are minimum and/or profits maximum.

22. **Cost of cultivation**: The expenditure incurred on all inputs and input services in raising a crop on a unit area is called cost of cultivation. It is expressed as rupees per hectare or rupees per acre.

23. **Cost of production**: The expenditure incurred in producing a unit quantity of output is known as cost of production, for example, Rs./kg of Rs./quintal.

24. **Independent variable**: Variable whose value does not depend on other variables and which influences the dependent variable, is termed as independent variable, for example, land, labour and capital.

25. **Dependent variable**: Variable whose value depends on other variables is termed as dependent variable, for example, crop output.

26. **Slope of a line**: It represents the rate of change in one variable that occurs, when another variable changes. Slope varies at different points on a curve but remains same on all points on a given line. It is the rate of change in the variable on vertical axis per unit change in the variable on horizontal axis and is expressed as a number.

27. **Total physical product**: Total amount of output obtained by using different units of inputs measured in physical units, for example, kg, tonnes, etc.

28. **Average physical product (APP)**: Output per unit of input on an average is termed as APP and is given by Y/X.

29. **Marginal physical product**: Addition to total output obtained by using the marginal unit of input & is measured as $\Delta Y/\Delta X$.

Farm production decisions involve application of the principles of production economics. Therefore, it is essential for a farm operator or a planner to understand basic concepts and relationships pertaining to the economics of agricultural production.

Production Economics

Meaning, Nature and Scope

Farm production economics is a broad division or field of specialization within the subject of agricultural economics. It is concerned with the choice of production patterns and resource use in order to maximize the objective function of the farm operator, their families, the society or the nation, within a frame work of limited resources. It is concerned with choosing of available alternatives or their combinations with a view to maximizing the returns and/or minimizing the costs.

Production economics is concerned with two broad categories of decisions in the production process:

i) How to organize resources in order to maximize the production of a single commodity *i.e.* to make choices from among various alternative ways of using resources.

ii) What combination of different commodities to produce.

The goals of agricultural production economics are two-fold:

i) To provide guidelines in using the farm resources most effectively

ii) To facilitate the most efficient use of resources from the stand point of the economy.

The Field of Agricultural Production Economics

1. With a view to optimizing the use of farm resources on an individual farm-level and to rationalize the use of agricultural resources from a national angle, production economics involves analysis of relationships and principles of rational decisions.

2. It is concerned with productivity *i.e.* use of and incomes from production resources. Specifically production economics includes

topics such as: combination of farm enterprises, methods of production, size of the farms, returns to scale, leasing, production possibilities, farming efficiency, use of credit and capital, risk and uncertainty which affect decision-making.

3. Any agricultural problem that falls under the scope of resource allocation and marginal productivity analysis, is the subject matter of agricultural production economics. The production economist is, therefore, concerned with any phenomena which have a bearing on economic efficiency in the use of agricultural resources.

The Objectives of Production Economics

The laws of production economics explain the conditions under which the quantities can be maximized (profit, output, national income), or minimized (cost, use of physical input). Therefore, the main objectives of production economics are:

i) To determine and define the conditions which provide for optimum use of resources,

ii) To determine the extent to which the existing use of resources deviates from the optimum use,

iii) To analyze the factors or forces which are responsible for the existing production patterns and resources use, and

iv) To delineate means and methods for changing the existing use of resources to the optimum level.

The principles or laws that help attain these objectives are the same on a micro as well as on a regional or national level. On micro level where intra-farm resource allocation and production pattern are involved, it is the subject matter of farm management.

Thus, while farm management is an intra-farm application of business principles, production economics involves inter-farm study and the economy of the region or the country as a whole.

Some of the frequently used terms in production economics analysis are:

Units of accounting: Application of inputs or measurement of output relate to a technical unit, plant or an economic unit.

Technical unit: Technical unit refers to a single, convenient unit in production for which technical coefficients are calculated *e.g.* an acre or a hectare of land, a cow, trio or a unit of poultry birds, *etc.*

Farm-firm: It is a production unit under one management and is also known as an economic unit referring to an aggregation of resources for which costs and returns are worked out as a whole. *e.g.* A farm holding is an economic unit.

Plant: Plant generally refers to a group of technical units, such as a dairy enterprise or say 15 acre farm.

Resources and Resource Services

There are some inputs or resources such as fertilizers (NPK), water, insecticides, which get consumed or transformed into products in the process of production. There are certain resources of which only services are available which are transformed into a product. Such resources are labour, implements, buildings, etc. Both resources and resource services are considered alike for our analysis that are transformed into the production of product. For example, labour and tractor fuel never enter physically into the transformation process, but we still consider services of these resources to be transformed to a product.

Fixed and Variable Resources

Level of some resources, such as buildings, machinery, implements is fixed over a planning period irrespective of the level of enterprise(s) taken up. These are known as **fixed farm resources**.

The resources, such as fertilizers, seeds, feeds etc. whose use varies with the level of enterprises are known as **variable resources**. Some fixed resources, however, have a variable use, such as tractor services, bullock power. This distinction between fixed and variable resources or inputs is essential for computing technical co-efficients for farm management decisions and production programming.

Flow and Stock Resources

There are some resources, if their services are not used, these cannot be stocked. Services are forthcoming like a flow. These services are labour, buildings, etc. If the services of a labourer are not used today, they cannot be stored until next day, month or the year. Such services are known as **flow services.**

Some resources such as seeds, fertilizer, feeds are entirely used up in the production process. If these resources are not used in one period of production, they can be stored for a later period. These resources are known as **stock resources.**

Some factors of production embody both flow and stock services, such as machinery. In this case, depreciation is a function of both use and time whether a service should be defined as a flow or a stock depends partly on the length of time under consideration. A building which lasts for 50 years provides a flow of services in each of the individual years. It gives flow service each year but a stock service over say 10 years (life of a tractor). Sunshine on the other hand is purely a flow service as are such other services derived from the atmosphere. The resources that give a flow service usually introduce elements of uncertainty in production planning.

Product or Production

Products are the result of the use of resources or services of resources: examples are wheat, maize, milk, eggs, etc. **Production** is a process of transformation of certain resources or inputs like land, labour (human), and bullock, seeds, fertilizer, irrigation water into products like wheat, paddy, milk.

Production function

Level of output of particular commodity depends upon the quantities of inputs used for its production. This relation between inputs and outputs can be characterized as a **production function**. **Production function** is, thus, a technical and mathematical relationship describing the manner and extent to which a particular product depends upon the quantities of inputs or services of inputs used. In the production function, output is "dependent upon" or "determined by" or "related to" or is the "function of" inputs or the use resources. Production function is of two types: (1) Continuous function (2) Discontinuous or discrete function

1. **Continuous function :** Continuous function can be explained by response of yield to fertilizer or seeds where the doses or levels of inputs and output can be split up into small units. e.g. Fertilizers can be applied to a hectare of land in quantities ranging from a fraction of a kilogram upto hundreds of kilograms. Labour is also quite divisible because it can be used in minutes or in hours.

2. **Discontinuous or discrete function :** Discontinuous or discrete function is obtained for input factors or work units which are used or done in whole members, such as one ploughing or a number of ploughings. One can only shift from one point to another. In discrete functions the alternative decision points get limited, otherwise same method of analysis is followed in both continuous and discrete production functions.

Transformation or Production Period : Time required for a resource to be completely transformed into a product is referred to as transformation period or the production period. The production period varies with resources. The production of the transformation period gives rise to complexities in decision-making. Economic analysis would be simple, if production is carried on with resources which are completely transformed in a single year. However, farmer must own long lived resources, such as machinery and buildings but these resources need heavy investments and decisions which extend over several years.

Short and long run production function

Production function which relates to factors and products, where some resources are fixed can be termed as **short run production function.** Those input-output relations which permit variation in the input of all factors (none is fixed) can be termed as **long run production function.**

A choice indicator : A choice indicator is a yardstick or an index or a criterion indicating which of two or more alternatives is optimum or will maximize a given end. The most desirable combination of products or factors cannot be determined without a choice indicator, which moves production problem from the field of physical sciences to the field of economics. e.g. Price ratios, substitution ratios, etc.,

Cost concepts : Corresponding to the fixed resources, there are costs which are there irrespective of the level of the enterprises or combination of enterprises. These are known as **fixed costs**. e.g. Farm building, machinery investment, etc.

On the other hand, there are **variable costs** corresponding to variable resources which vary with the level of enterprises or production. e.g. seed, feed, fertilizer costs. **Fixed and variable costs put together amount to total cost of production.**

Return concepts : Gross returns are equal to total production times the price. Returns to fixed farm resources farm resources are equal to gross returns minus variable costs. These are also known as **returns over variable costs**. Net returns are equal to gross returns minus all costs (fixed and variable: cash and kind).

Returns over variable costs = Gross returns – Variable costs

Net returns = Gross returns – all costs

Farm entrepreneur : Farm entrepreneur is the person (also a group) who thinks of, organizes and operates the business and is responsible for the results. i.e. losses and gains from the business. He is a pioneer in organizing and developing the farn-firm.

Farm Manager : Farm Manager is a person who manages or supervises the business according to the instructions of the entrepreneur. He is hired to manage the business and he is not generally responsible for any gain or loss to the business.

Farm Executive : Farm Executive carries out the orders of the management. **A farmer may be performing all the above three functions himself.**

Agricultural Production Economics vis-a-vis Farm Management

Following are the differences between agricultural production economics and farm management :

S.No	Agricultural Production Economics	Farm Management
1.	It is a science in which the principles of choice are applied to use of land, capital, labour and management of resources in the farming industry.	It is a science of organization and operation of farm with a view to earn continuous profits.
2.	Agricultural production economics is a specialized branch of agricultural economics.	It is an integral part of agricultural production economics.
3.	It is microeconomic in its scope as it deals with the problems of farming industry.	It is microeconomi c in its scope as it is concerned with the problems of individual farm.
4.	It deals with allocative efficiency of the use of resources in agriculture.	It deals with economics efficiency at the farm level.
5.	It is an inter-farm study.	It is an intra-farm study.

❑❑❑

Chapter 3

Production and Production Functions Production Relations (F-P, F-F and P-P Relationships)

Production of farm commodities involves numerous relationships between resources and products. Some of these relationships are simple, others are complex. Knowledge of these relationships is essential, as they provide the tools by means of which the problems of production or resource use can be analyzed.

Major production relationships are :

1. Factor -Product relationship
2. Factor -Factor relationship
3. Product-Product relationship

1. Factor-Product Relationship:

1. It deals with the **production efficiency** of resources.
2. The rate at which the factors are transformed into products is the study of this relationship.
3. **Optimization of production** is the goal of this relationship.
4. This relationship is known as input-output relationship by farm management specialists and fertilizer responsive curve by agronomists.
5. Factor -Product relationship guides the producer in making the decision **'how much to produce?'**.
6. This relationship helps the producer in the determination of optimum input to use and optimum output to produce.

7. **Price ratio** is the choice indicator.
8. This relationship is explained by the **law of diminishing returns** (e.g. Additional fertilizer dose has a negative effect on yeild).
9. Algebraically, this relationship can be expressed as Y = f (X1 / X 2,X3.................Xn)

2. Factor-Factor Relationship

1. This relationship deals with the resource combination and resource substitution.
2. **Cost minimization** is the goal of factor -factor relationship.
3. Under factor-factor relationship, output is kept constant, input is varied in quantity.
4. This relationship guides the producer in deciding **'How to produce'**.
5. This relationship is explained by **the principle of factor substitution or principle of substitution between inputs.**
6. Factor-Factor relationship is concerned with the **determination of least cost combination of resources** (e.g. Bullocks or tractor, harvest by machinery or labour, handmilking or by machines).
7. The choice indicators are **substitution ratio and price ratio.**
8. Algebraically, it is expressed as Y = f(X1 X2 / X3 X4 ..... Xn)

In the production, inputs are substitutable. Capital can be substituted for labour and vice versa, grain can be substituted for fodder and vice versa. The producer has to choose that input or inputs, practice or practices which produce a given output with minimum cost. The producer aims at cost minimization i.e., choice of inputs and their combinations.

Isoquants

The relationship between two factors and output cannot be presented with two dimensional graph. This involves three variables and can be presented in three dimensional diagram giving a production surface. An isoquant is a convenient method for compressing three dimensional picture of production into two dimensions.

Definition : An isoquant represents all possible combinations of two resources (X1 and X2) physically capable of producing the same quantity of output. Isoquants are also known as isoproduct curves or equal product curves or product indifference curves.

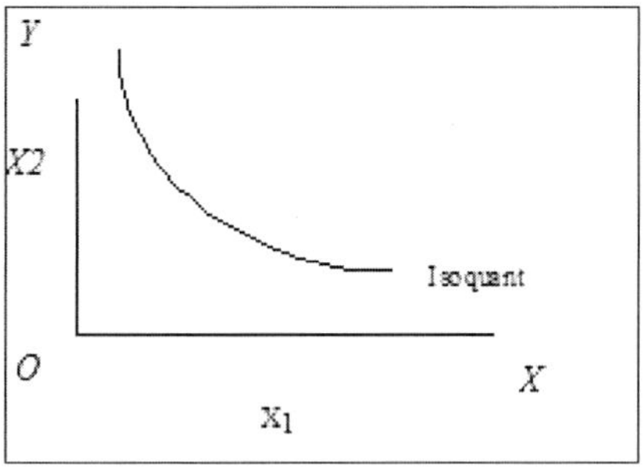

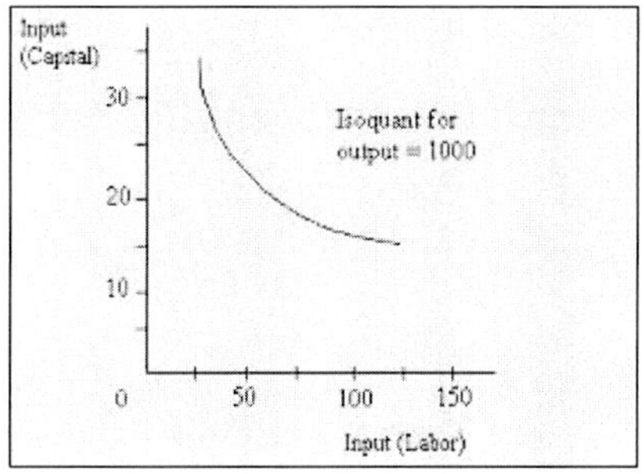

Iso Revenue Line (IRL) : Iso revenue line is defined as "a line depicting all combinations of two products that will generate a given (or same) level of total revenue". It represents the rate at which the market is willing to exchange one product for another.

IRL is the locus of points of various combination of quantities of y and x whose sale yields the same revenue to the firm. The slope of the iso revenue curve is equal to the ratio of the prices of the commodities Slope of iso revenue = OA / OB = P_x / P_y

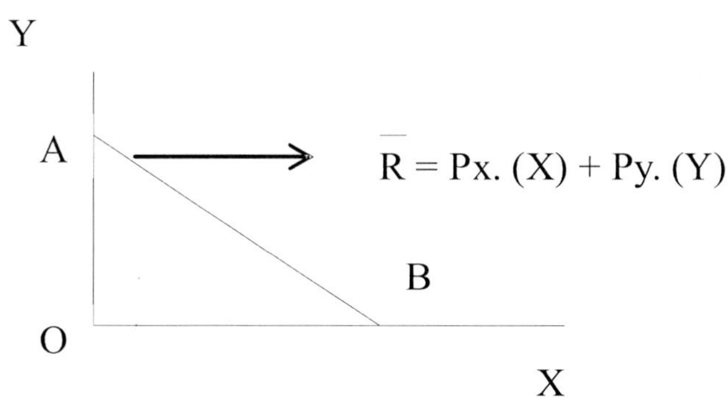

$\bar{R}$ = Total revenue

(a) Y only sold only - total revenue is OA

Py = $\bar{R}$ i.e. the quantity of y which yields

$\bar{R}$ is (OA) = $\bar{R}$ / P_y

(b) For x $\bar{R}$ = is OB = R / P_x

The choice indicator for a farm Product – Product relationship is provided by an iso revenue curve. The iso revenue line indicates the ratio of prices for the two competing products. It is a line which defines all the possible combinations of two commodities, which would yield an equal revenue or income. The line AB is an iso revenue line.

3. Product-Product Relationship

1. Product-Product relationship deals with resource allocation among competing enterprises.
2. The goal of Product-Product relationship is **profit maximization**.
3. Under Product-Product relationship, inputs are kept constant while products (outputs) are varied.
4. This relationship guides the producer in deciding **'What to produce'**
5. This relationship is explained by the **principle of product substitution and law of equimarginal returns** (Eg. Best allocation of resources/fertilizers).
6. This relationship is concerned with the **determination of optimum combination of products (enterprises).**
7. The choice indicators are **substitution ratio and price ratio.**
8. Algebraically it is expressed as Y1=f (Y2 Y3, ... Yn)

Production Possibility Curve (PPC)

Production Possibility Curve is a convenient device for depicting two production functions on a single graph.

Def: Production Possibility Curve represents all possible combinations of two products that could be produced with given amounts of inputs.

Production Possibility Curve is known as Opportunity Curve because it represents all production possibilities or opportunities available with limited resources.

It is called Isoresouce Curve or Iso factor curve because each output combination on this curve has the same resource requirement.

It is also called Transformation curve, as it indicates the rate of transformation of one product into another.

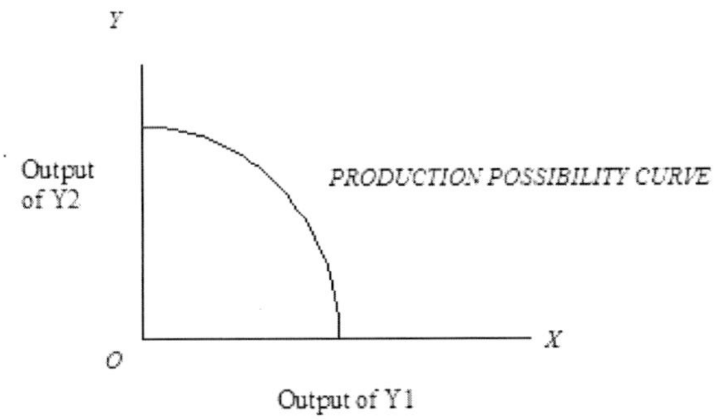

Enterprise Relationship and their Combination

Types of Product-Product Relationships or Enterprise Relationship

Farm commodities bear several physical relationships to one another. These basic product relationships are as following :

The enterprise physical relationships may be :

1. Joint Product enterprises
   a) Fixed proportions
   b) Variable proportions
2. Competitive enterprises with
   a) Constant rate of substitution
   b) Increasing rates of substitution
   c) Decreasing rates of substitution
   d) Other combinations of independent products
3. Complementary enterprises
4. Supplementary enterprises
5. Antagonistic products

1. Joint Product enterprises

Joint products are produced through a single production process and one of the products cannot be produced alone but must be accompanied by one or more of other products.

e.g. *Ceiba pentandra* -silk cotton lint and cotton seed, Paddy and straw, etc.

2. Competitive enterprises

Two enterprises are competitive in the use of given resources, it output of one can be increased only through a sacrifice in the production of the other. The nature of enterprise relationship (Production Possibility Curve) is dependent upon the nature of production function for each independent enterprise.

This relationship holds good, when increase or decrease in the production of one product affects the production of the other commodity inversely.

e.g.1. Drumstick and curryleaf can be grown in teak based agroforestry system grown. Arecanut with pepper, banana, cardamom, cocoa, vanilla, etc.

2. Fruit trees like mango, sapota, tamarind, etc. can be grown in any short rotation tree species. i.e. Competitive enterprises compete for farm resources and substitute for each other.

3. Complementary enterprises

Two products are said to be in complementary relationship, when an increase in output of one results in an increase in output of the other, with resource amount held constant.

In other words, a shift of resources from a first crop to a second crop, will increase rather than decrease the output of the first.

e.g. 1. Sandal tree species with host tree sps

2. If maize is grown after berseem crop (*Trifolium alexandrinum*), the yield of maize increases because berseem crop fixes nitrogen and makes the soil more fertile for the next crop.

4. Supplementary enterprises

Two enterprises are in supplementary relationship, when the output of one product can be increased with neither a gain nor a sacrifice in another product, with resources constant.

This relationship exists when increase or decrease in the level of one product doesn't affect the production level of the other product.

e.g. Wheat and maize crops in relation to land.

5. Antagonistic products

Antagonism is expressed when the production function for two independent products changes, as the two enterprises are produced in the presence of each other. Antagonism is the opposite complementary relationship.

Two products may be detrimental (harmful) to each other because of disease or similar factors. When this is true, only one of the products should be produced. e.g. Aquaculture and paddy cultivation.

❑❑❑

Chapter 4

Principles Involved in Forest Business Management

Principles Involved in Farm Management Decisions

Farm Management principles serve as a guideline for collecting and using requisite information for rational decision making. They also provide a set of tools for the preparation of farm budgets and production programmes. These principles help provide answers to various farm problems and save time and energy. So, knowledge of these principles enhances farm entrepreneurs' sense of judgment, a pre-requisite for meeting the demands of a business under odd situations.

The following are the six basic principles involved in making rational farm management decisions :

I. Principle of variable proportions or laws of returns

II. Principle of Costs

III. Principle of substitution between inputs

IV. Principle of Equi-marginal returns or Principle of opportunity cost

V. Principle of substitution between products

VI. Principle underlying decisions involving time and uncertainty

These farm management principles are briefly discussed below:

I. Principle of variable proportions or laws of returns

It has three phases :

a) Diminishing returns

b) Constant returns

c) Increasing returns

a) **Diminishing returns :** It is a basic natural law affecting many phases of management of a farm business. It is a law of fundamental importance in agriculture. This law describes the relationship between output and a variable input, when other inputs are held constant. **The law can be stated as follows : "If increasing amounts of one input are added to a production process while all other inputs are held constant, the amount of output added per unit of variable input will eventually start decreasing."** It states that if the quantity of one factor is increased with quantities of other factors held constant, the marginal increment to the total product may increase or remain constant at first but will eventually decrease after a certain point.

Law of Diminishing Marginal Returns (LDMR): The principle that if technology is unchanged, as more units of a variable resource are combined with one or more fixed resources, the marginal product of the variable resource must eventually decline.

The Law of Diminishing Marginal Returns (LDMR)

- Total Product (TP) - This is the total output produced by workers
- Marginal Product (MP) -This is the output produced by an extra worker

Diminishing Returns occurs in the short run, when one factor is fixed (e.g. Capital). If the variable factor of production is increased, there comes a point where it will become less productive and therefore, there will eventually be a decreasing marginal and then average product. This is because if capital is fixed extra workers will eventually get in each other's way, as they attempt to increase production. e.g. think about the effectiveness of extra workers in a small café. If more workers are employed production could increase but more and more slowly. This law only applies in the short run because in the long run all factors are variable.

Assume the wage rate is £10, then an extra worker Costs £10. The Marginal Cost (MC) of a sandwich will be the Cost of the worker divided by the number of extra sandwiches that are produced.

Therefore, as MP increases MC declines and vice versa

A good example of diminishing returns includes the use of chemical fertilizers- a small quantity leads to a big increase in output. However, increasing its use further may lead to declining Marginal Product (MP) as the efficacy of the chemical declines.

Diagram of Diminishing Returns

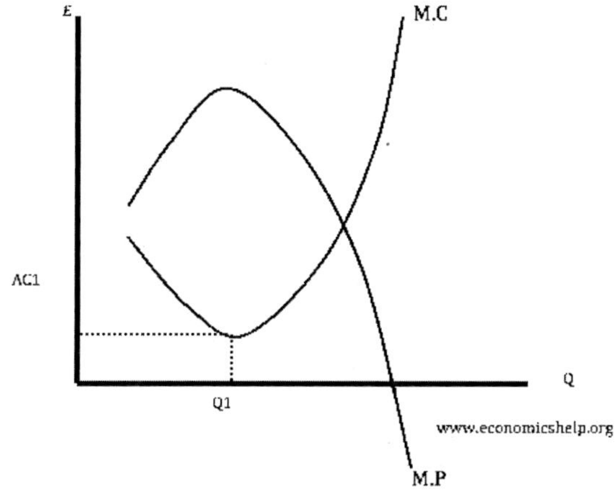

b) **Constant returns :** It means each marginal unit of a variable resource adds the same amount of the output to the total production. Though, **"diminishing marginal productivity"** is the rule, constant productivity is frequently observed, when no resource is fixed and all are increased together in the same proportion. **For example**, another acre may be as productive as the first with same inputs. If one acre of wheat requires 20 man-hours of labour, 30 kgs of seed and 13 inches of irrigation water and yields 10 quintals of wheat, the second acre will require an additional 20 man-hours of labour, 30 kgs of seed and 13 inches of irrigation water and will also yield 10 quintals of wheat. The second acre is just as productive. The marginal or added production from each increase in resource input is the same; this is a case of **constant productivity.**

c) **Increasing Returns :** Increasing productivity means added resources give increasing returns. This relationship may hold only over a very limited range of production and is applicable, when all the resources are increased together and not when some resources are fixed. For example, a cattle shed constructed for 30 cows may cost more per cow, then if one constructed for 60 cows, the cost involved in the latter case may not be double because of some economics on account of joint walls, etc., but the gross returns per cow might be the same. Use of added resources thus, will give increasing returns in such cases. In this case each additional unit gives higher and higher returns. So long as this relationship holds, production should keep expanding.

II. Principle of Cost

1. This principle guides the producers in the minimization of losses. Costs are divided into fixed and variable costs. Variable costs are important in determining whether to produce or not. Fixed costs are important in making decisions on different practices and different amounts of production.
2. In the short run, the gross returns or total revenue must cover the total variable costs (TVC). To state in a different way that selling price must cover the average variable cost (AVC) to continue production in the short run.
3. In the long run, gross returns or total revenue must cover the total cost (TC). Alternatively stated, that the selling price must cover cost of production (ATC).
4. In the short run MR = MC point may be at a level of output which may involve loss instead of profit. The situation of operating the farms, when the price of product (MR) is less than average total cost (ATC) but greater than average variable cost (AVC) is common in agriculture. This explains why the farmers keep farming even when they run into losses.
5. The maximum net revenue is obtained, when marginal cost (MC) equals the price of the product (MR)

III. Principle of substitution between inputs

1. This economic principle explains one of the basic production relationships viz., factor factor relationship. It guides in the determination of least cost combination of resources. It helps in making a management decision of how to produce.
2. Substitution of one input for another input occurs frequently in agricultural production. For example, one grain can be substituted for another or forage for grain in livestock ration, chemical fertilizers can be substituted for organic manure, machinery for labour, herbicides for mechanical cultivation, etc.
3. The farmer must select that combination of inputs or practices which will produce a given amount of output for the least cost.
4. In other words, the problem is to find the least cost combination of resources, as this will maximize profit from producing a given amount of output.

e.g. Bullocks or tractor, harvesting of crops by machines of labour, milking machines or handmilking.

IV. Principle of Equi-marginal returns or Principle of opportunity cost

1. Most of the farmers have limited resources. They have limited land, limited capital, limited irrigation facilities. Even the labour which is considered to be surplus becomes scarce during peak sowing, weeding and harvesting periods.
2. Under such resource limitations, farmers must decide how a limited amount of input should be allocated or divided among many possible uses or alternatives.
3. For example, farmer has to decide on the best allocation of fertilizer between different crops and feed between different types of livestock. In addition, limited capital must be allocated to the purchase of fertilizers, seeds, feed, etc.
4. The equi-marginal principle provides guidelines for the rational allocation of scare resources. The principle says that returns from the limited resources will be maximum, if each unit of the resource should be used, where it brings greatest marginal returns.
5. This law states that "a limited input should be allocated among alternative uses in such a way that the marginal value products of the last unit are equal in all its uses".

V. Principle of substitution between products

1. This principle explains the product-product relationship and helps in deciding the optimum combination of products. Also, this economic principle guides in making a decision of what to produce.
2. It is economical to substitute one product for another product, if the decrease in returns from the product being replaced is less than the increase in returns from the product being added.
3. The principle of product substitution says that we should go on increasing the output of a product so long as decrease in the returns from the product being replaced is less than the increase in the returns from the product being added.

VI. Principle underlying decisions involving time and uncertainty

Farm management involves dynamic adjustment of the farm organization and operations. Such an adjustment relates to taking account of :

a) Time element in the calculation of present value of future incomes, and

b) The risks and uncertainties involved in farm operations over time e.g. Natural calamities and fluctuation in price of input and output.

Definitions

1. **Law Of Diminishing Marginal Returns (LDMR) :** The principle that if technology is unchanged, as more units of a variable resource are combined with one or more fixed resources, the marginal product of the variable resource must eventually decline.
2. **Marginal Product (MP) :** The change in output that results from changing labour input by one unit. **MP = "Q/"L**
3. **Marginal Revenue Product (MRPL) :** The change in the total revenue that results from changing labour input by one unit. MRPL = "TR/"L
4. **Gross Substitutes:** Inputs such that when the price of one changes, the demand for the other changes in the same direction because the substitution effect exceeds the output effect.
5. **Gross Complements :** Inputs such that when the price of one changes, the demand for the other changes in the opposite direction because the output effect exceeds the substitution effect.
6. **Own-wage Elasticity of Demand (eD) :** A measure of the responsiveness of the quantity of labour demanded to a change in the wage rate.

   $eD = - \%\Delta L / \%\Delta w$
7. **Cross Elasticity of Demand:** A measure of the responsiveness of the quantity demanded of input i to the change in the wage of input j.

   $eijD = \%\Delta Li / \%\Delta wj$

   a) eijD > 0 inputs i and j are gross substitutes.

   b) eijD < 0 inputs i and j are gross complements.

8. **Total Wage Bill:** The total wage cost to the firm; the wage rate multiplied by the quantity of labour hours employed.
9. **Total Wage Bill Rules:** Rules for determining the elasticity of labour demand. Labour demand is elastic, if a change in the wage rate causes the total wage bill to move in the opposite direction. Labour demand is inelastic, if a change in the wage rate causes the total wage bill to move in the same direction.
10. **Determinants of Elasticity of Demand for Labour:** These include: product demand, ease of substituting other inputs, the elasticity of supply of other inputs, and the share of labour cost in the total costs of the firm.
11. **Minimum Wage:** A wage floor where a legally established minimum rate of pay is specified for labour employed in covered occupations.
12. **General Training:** Skills that can enhance a worker's productivity with a wide variety of employers.
13. **Specific Training:** Skills that can enhance a worker's productivity with only one employer.
14. **Human Capital:** The accumulation of prior investment in education, OJT, health and other factors that increase productivity.
15. **Screening Hypothesis:** The view that education only identifies individuals who are trainable or of high ability rather than increasing productivity per se.
16. **Private Perspective (for Investment in Human Capital):** Investors consider only those costs and benefits accruing to themselves when deciding whether to invest in human capital or not.
17. **Social Perspective (for Investment in Human Capital):** Because education produces substantial external or social benefits which accrue to everyone in society, education would be under-produced, if these benefits were ignored in the provision of education.
18. **Capital Market Imperfections**: The bias against lending money for investments in human capital that occurs largely because human beings cannot be used as collateral for loans.

19. **Marginal Rate of Substitution (MRS):** In economics, the marginal rate of substitution is the rate at which a consumer is ready to give up one good in exchange for another good while maintaining the same level of utility. At equilibrium consumption levels (assuming no externalities), marginal rates of substitution are identical.

❑❑❑

Chapter 5

Forest Business Analysis: Advantages of Forest Business Records and Accounts

The basis of progress and success in business is experience. It is essential for students to understand the necessity and importance of maintaining farm records:

1. To enhance their ability to handle a set of records and accounts;
2. To improve the analytical ability in processing the data collected in farm record books; and
3. To improve the competency in interpretation of the results.

Farm business analysis is dealt with under different names *viz.*, Farm Accountancy, Farm Records and Accounts, or Farm Book-keeping.

Farm Accountancy : Farm Accountancy is defined as the art as well as the science of recording in books business transactions in a regular and systematic manner so that their nature, extent and financial effects can be readily ascertained at any time of the year.

Farm Book keeping : Farm Book-keeping is known as a system of records written to furnish a history of the business transactions, with the special reference to its financial side.

Farm Business Analysis

The main object of farm business analysis is to examine as to:

i) how does the business fare at a certain time

ii) where are the weaknesses

iii) what improvements are possible

There are some subsidiary objectives, such as providing background information for farm policies and for getting credit policies, etc. The following are the three steps or stages of farm business analysis.

i) Proper recording of accounts and activities
ii) Analysis and interpretation of results
iii) Presentation of results

Advantages of Forest Records and Accounts

The various advantages of keeping systematic forest records can be described as under:

1. Means to higher income
2. Basis for diagnosis and planning
3. Way to improve managerial ability of the farmer
4. Basis for credit acquisition and management
5. Guide to better home management
6. Basis of conducting research in Forest Economics and production economics
7. Basis for government policies

1. **Means to higher income :** To obtain higher income, farmers must have exact knowledge about present and potential gross income and operating costs. The best way to obtain information on present results is to keep records and accounts, in order to:
   i) To know the financial status at a point of time
   ii) To know gains and losses over time
   iii) To know the better sources of income and items of cost
   iv) To keep a check on unproductive expenditure
   v) To examine comparative profitability and costs involved for different enterprises
   vi) To locate weak points in the farm organization
   vii) To compare farm efficiency with the farmers operating under comparable farming situations
   viii) To develop rational short-term and longterm production plans.
2. **Basis for diagnosis and planning :** Diagnosis of management problems is the pre-requisite of sound planning. Records and accounts provide the basic information needed for such a diagnosis.

3. **Way to improve managerial ability of the farmer :** It helps to acquire business habits which can be of help in taking advantage of changes in the economic environment. The farmer gets a better insight into the working of his business, which helps him in finding out the defects which can be set right by exercising better control and effecting economics. The farmer can avoid mistakes and losses which would otherwise result due to dependence only on his memory for guidance.
4. **Basis for credit acquisition and management :** Properly kept record and accounts can demonstrate and authenticate the production and income potentials and credit worthiness of the farmer. Thus, lending agencies can help the farmer in meeting his credit needs more readily and he can manage his credit utilization properly.
5. **Guide to better home management :** Records and accounts provide information on farm-household economy. This is particularly important in India, where farm and home management are so closely integrated. Analysis of farm records provides good guides for the allocation of resources between production improvement and immediate family welfare.
6. **Basis of conducting research in Forest Economics and production economics :** Research requires precise and correct data which is possible only if proper records and accounts are maintained on the forest included in the study.
7. **Basis for government policies :** The farmers need to continuously feed the facts for state and national forest policies, such as land policies and crop insurance, etc. Record and accounts are helpful in obtaining the correct data for examining and developing such policies to be sound.

Problems and Difficulties in Forest Accounting

Most of the Indian farmers have distaste for maintaining farm/ forest account due to lack of education, business orientation, and time required to do this job. Some of the specific difficulties in maintaining accounts are:

1. **Subsistence nature of farming :** Farming as a business is a small size operation in India. Farmers cannot engage separately trained

accountants for helping them in farm accounting. Subsistence nature of farming does not produce any incentive for keeping the records.

2. **Farming is a laborious work :** Farming requires a lot of physical labour, in addition to mental work of management. In the daily routine, the farmer usually gets exhausted in the evening and does not feel like sitting at the desk to complete records and accounts.

3. **Triple role of Indian farmer :** Indian farmer plays a triple role in running his business, i.e. that of a manager, a financier and a labourer. Therefore, he needs to know his wages as a labourer, his returns to capital, and his returns to his management role. Complex type of records which would give such information are difficult to maintain.

4. **Illiteracy and lack of business awareness :** The very low level of illiteracy among the Indian farmers is a hindrance in developing the required level of business awareness on the part of the cultivators and therefore, they do not realize the need for records and accounts.

5. **Complicated nature of the agriculture business :** It is a biological industry and is always subject to weather and other natural uncertainties. It requires an accounting system which can handle various complexities involved in the business of farming. Such complicated accounts are difficult to maintain.

6. **Inadequate extension service : S**ufficient number of trained specialists in farm management is not available who could help farmers maintain accounts of their business. Recently Punjab Agricultural University, Ludhiana has taken a lead in India, in appointing Assistant Extension Specialist (Farm Management) at each district headquarters. One of their functions is to help farmers maintain farm records and accounts. They have started assisting farmers to maintain systematic records and accounts on demonstration centres in each district and that needs to be further strengthened. The organization of farm records association will help their members-farmers in maintaining accounts, analyzing the data and interpretation of results.

7. **Non-availability of suitable farm record books :** Lack of standardized, easy to understand and maintain account books

or proforma also stand in the way of willingness of the cultivators to keep records. Standard farm record books need to be developed which may be simple and easy to understand and available in local languages. Punjab Agricultural University has designed a very simple record book to help farmers.

8. **Fear of taxation :** Farmers are always afraid of new taxes. They fear that if they maintain records and accounts and their incomes show up high, some sort of tax may be levied on them.

Accounting period

It may be desirable to consider an accounting period that fits into the production period. India has distinct seasons of production, so agricultural year may be more suitable and convenient accounting period for most parts of the country, i.e. starting from 1st July and ending 30th June of next calendar year.

Levels of Farm Accountancy : A good system accounts for any farm is one that enables recording of information that the farmer needs and also permits the desired analysis of the information recorded.

Farm records are usually of the following types :

i) Farm inventory

ii) Farm Financial Accounting

iii) Farm Cost Accounting : (a) Full cost accounting (b) Single Enterprise Accounting.

Financial Accounting : These records and accounts relate to the operations and activities of the farm business as a whole. Records and accounts under this method set forth income, expenditure, and profit or loss of the farm business as a whole. They reveal the aggregate effect of all the farm undertakings, such as crops growing, dairy, poultry, etc. It gives an overall picture of the entire business and is simpler than cost accounting.

Cost Accounting

a) **Full cost accounting** : Under this, accounts show income and expenditure and profit or loss of the farm business as a whole as well as of each department, each farm product or commodity as a separate unit. An account is kept for each enterprise or department and all income and expenditure pertaining to the individual farm enterprise is credited or debited to its respective account. This method is useful especially for large farmholders who can afford the necessary labour or money for keeping such accounts.

b) **Single Enterprise Accounting :** Accounts can be kept for important selected enterprises or a particular aspect of an enterprise, such as production of milk, sugarcane, wool, eggs, etc. This is only a restricted from of cost accounting. It is a simpler method and takes much less time but is not so accurate and complete.

Systems of Book Keeping

There are two systems of farm accountancy: (i) double entry system (ii) single entry system

i) **Double entry system :** It is a method of recording each transaction in the books of accounts in its two-fold aspects, i.e. two entries are made for each transaction in the same set of books, one being a debit entry and the other a credit entry.

Theory of double entry : Every business transaction involves two parties - one for receiving the goods or services and the other for giving them. Therefore, every transaction is entered at two places, for credit and for debit. The following two examples from an agroforestry system will illustrate the point:

1. **Sale of wheat (intercrop) for Rs. 240/- :** Two accounts involved will be cash account and wheat account. The cash account will be the receiving account and hence, the amount will be written on the debit side of it. The wheat account will be the giving account, hence the amount will be written on the credit side of it.

2. **Receipt of money from Mr. Singh Rs. 250/- :** Here Rs. 250/- have again been received in the cash account. The cash account is, therefore, debtor for the sum received. Mr. Singh has paid the amount and therefore, Mr. Singh's account will be the creditor. Cash account will be debited and Mr. Singh's account will be credited.

The Advantages of Double Entry Systems are as under :

1. The system provides for complete personal and impersonal records which include assets, gains or losses. Thus, nature and value of the possessions can be ascertained.
2. It furnishes ways and means for checking arithmetical accuracy because two entries are made for each transaction.
3. The system provides detailed information regarding the business, classified records of all transactions will show combined overall results of a given policy.
4. The system provides for automatic checks to prevent mistakes.
5. It affords an easy and ready reference to details of accounts.

ii) **Single Entry System :** This is the system which ignores the double effect of transactions. Only personal accounts of debtors and creditors are kept and impersonal accounts are ignored altogether. In fact, any method of accounting which falls short of double entry, may be called as single entry system. Therefore, it is relatively imperfect and its results are less reliable and its accuracy cannot be tested by means of a trial balance, which is possible under the double entry system alone.

A good forest record system should :

i) Be easy to keep

ii) Give needed information

iii) Provide the information when it is needed, and

iv) Permit analysis of the information recorded.

Types of Forest Business Records and Accounts (Forest Record Books)

There are many different kinds of farm records and accounts, each of which can be adopted for a given purpose on a particular farm situation. It is important to understand the basic principles for the preparation of forms and proforma of farm records and qualities of a good farm record system. They vary with the purpose, system of farming, level of understanding and capability of the farmer.

The following are the main characteristics of a good farm record book :

1. Simple and easy to understand
2. Suitable forms for recording the type of information the farmer wants to record
3. Provision for an itemization and classification of all entries
4. Adequate space for itemizing all entries, and the lines and spaces sufficiently wide for writing without crowding.
5. Ample space in each section for all the entries that are likely to be recorded
6. Adequate instructions for recording and analysis of the recorded data.

Part of a Farm Record System

There are three parts of a farm record system :

i) Physical Farm Records
ii) Financial Farm Records
iii) Supplementary Farm Records

i) **Physical Farm Records :** Physical records are related to the physical aspects of the operation of a farm business. They do not indicate financial position or the outcome of the farm business, but simply record the physical efficiency or performance of the farm.

Physical farm records normally include the following records :

1. Farm map, soil map and contour map
2. Charts on physical efficiency
3. Land utilization records
4. Crop production and disposal record
5. Livestock production and disposal record
6. Labour records, daily work diary
7. Machinery use records

8. Feed records
9. Stock/store register
10. Poultry records

1. **Farm map, soil map and contour map :** It is one of the most simple and brief record of farm features which gives a visual impression or a birds' eye view of the whole location of the farm business. Personal observation is very necessary and the farm map is just supplementary to the physical spot observation.

   e.g. A farm layout map should contain farm boundary, buildings, tubewell, water channels, approach road & main road.

ii) **Financial Farm Records :** In order to provide information regarding the profitability of the whole farm business over a given period, the financial records are important to be maintained. They enable the analysis to be carried out to reveal the economic strengths and weaknesses of the farming system, and to provide data to help in the preparation of revised plans and budgets.

The following financial records should be kept :

1. Farm inventory
2. Farm cash or farm financial record
3. Classified farm cash account and annual business analysis
4. Capital asset and sale register
5. Cash sale register
6. Credit sale register
7. Purchase register
8. Wage register
9. Funds borrowed and repayment register
10. Farm expenses paid in kind register
11. Non-farm income record

iii) **Supplementary Farm Records :** The following Supplementary Farm Records should be kept :

1. Sanction register
2. Auction register
3. Rainfall register
4. Hire register
5. Stationary register

Methods of Valuation

The use of a particular method of valuation will depend upon the purpose of the valuation and nature of the assets being valuated. Following are the commonly used valuation methods:

1. Cost minus depreciation
2. Cost or market price, whichever is lower
3. Net selling price
4. Replacement cost minus depreciation
5. Income capitalization

1. **Valuation at Cost Minus Depreciation :** With cost as the basis of valuation, the inventories show the total of sums actually put into the business and amount depreciated over time. This method is commonly used for such working assets as machinery and breeding livestock. *This method assumes that the purchase price was an approximation of the value of the asset and its value in subsequent years can be determined by subtracting a depreciation allowance from its cost.* By using a systematic method of charging depreciation to the accounting periods, financial position is reasonably well estimated and the income figure is not distorted by the market fluctuations. This method cannot be, however, applied to things produced on the farm.
2. **Valuation at Cost or Market Price (whichever is lower):** *In this method valuation is estimated at the cost or the market price, whichever is lower.* This method is commonly used for valuing purchased farm supplies. No paper profits accrue and losses due to falling prices are absorbed immediately.
3. **Valuation at Net Selling price:** *This means the price which could probably be obtained for the asset if marketed, less the cost of marketing.*

This conforms most closely to the present worth. This method, is used for those items that are held primarily for sale. This method, which represents the market price (less the selling cost) provides a reasonably accurate measure of the current value of the asset. This method may be usefully applied to crops or the livestock produced for the market.

4. **Valuation by Replacement Cost Minus Depreciation:** *This method is to valuate the assets at what it would cost to reproduce them at present prices and under present methods of production.* This method is best suited for long lived assets, such as buildings particularly where wide changes in the price level occur. This method will guard against undervaluation, but may not ensure against over valuation.

5. **Valuation by Income capitalization method:** *This method is appropriate for the farm assets whose contribution to the income of the farm business can be measured and which have a long life.* The *capitalization formula V=R/r can be used for this purpose, where "V" is the value in rupees, "R" refers to the constant income over infinite number of years in future, and "r" is the rate of interest.* This method can be used to valuate the assets like land. In practice neither the annual income nor the interest rate in future is known with accuracy. As a result, this method is generally used in combination with other methods, such as the market price.

Selection of the Method of Valuation

Following are some of the suggestions, which can be useful in valuation of the farm assets :

1. All assets that will be sold within the year can be valued at the net selling price.
2. All supplies can be valued at the cost or the market price (whichever is lower).
3. Working capital assets, such as machinery and equipment can be valued by use of cost minus depreciation method.
4. The farm buildings that have been constructed a long time ago can be valued by replacement cost minus depreciation method. The farm buildings that have been constructed only a short time ago may be valued at cost minus depreciation.

5. Farmland may be valued with capitalization formula in combination with market price.

It is, therefore, necessary to understand the various methods of calculating depreation.

❑❑❑

Chapter 6

Methods of Computing Depreciation and Net Worth Statements

Depreciation is a slow using up of a long lived asset. It represents the amount by which a farm resource decreases in value as a result of cause other than a change in the general price of the item. In other words, it is a decline in the value of a given asset as a result of the use, wear and tear, accidental damage and time obsolescence. Depreciation involves prorating the original cost of an asset over its useful life. The amount of depreciation charged should, therefore, correspond to the loss in the value of the asset over time. The computation of depreciation would not be necessary, if all the items purchased were completely worn out by the end of each year. However, the items, such as buildings, equipment and livestock, etc. are used up gradually over a long period of years and an important question arises about the determination of the cost of such articles for one specific year. Depreciation occurs because of factors like ordinary wear and tear in use, complete exhaustion of the article, unusual damage or deterioration and obsolescence. The value of the asset may become completely exhausted or reduced to its junk value.

Method of Computing Depreciation : There are many methods of computation of depreciation but none of these methods can be considered the most appropriate under all circumstances. The methods of computing depreciation can be based on either of the following two assumptions:

i) Assets are used on a constant rate year after year;

ii) Assets are used at varying rates year after year.

The choice of the method mostly depends upon the rate at which the asset loses its value.

The most common methods of computing depreciation are :

1. Annual revaluation
2. Straight line method
3. Diminishing balance method
4. Sum of the year digits method or Reducing fraction method
5. Compound Interest method

1. Annual revaluation

This method implies estimating the market value of the asset in the beginning and the end year inventory and then taking the difference as depreciation.

This method has the limitation that estimation of establishing annual values for items, such as machines and buildings that are not bought and sold frequently becomes difficult. Also, mere changes in prices may render this method unrealistic. This method is, however, useful in case of livestock in the early years of life, i.e. in the appreciating phase.

2. Straight line Method

This method is easy, simply and usually very satisfactory for most purposes. This method assumes that assets are used more or less to the same extent every year and therefore, equal amounts of costs on account of their use can be charged every year.

This method consists of dividing the total anticipated depreciation by the number of years the particular asset is expected is expected last. The total depreciation is the purchased price of the article minus junk value, salvage value, scrap value or residual value, if any.

By this method, depreciation will be

$$= \frac{\text{Purchase Price of the Asset - The junk value}}{\text{No. of useful years of life (expected life)}}$$

Suppose a tractor costs Rs. 24,000, when new and expected to last 10 years. The junk value of the tractor after 10 years would be Rs. 4,000/-. Then the annual depreciation would be :

$$= \frac{24{,}000\text{-}4000}{10} = Rs.2000/\,year$$

It is advisable not to charge the junk value of the most of the farm assets because it is usually very negligible. Information on the number of useful years of a particular asset can be obtained from Engineers. In case of a second hand machine, one can make personal estimates. This method does not suit where assets get used up at varying rates, e.g. A tractor for example depreciates much more during first few years than in later years.

3. Diminishing balance method

A fixed rate of depreciation is used for every year and applied to the value of the asset at the beginning of the year.

Depreciation can be computed by establishing a constant percentage rate of depreciation and using this rate each year to the remaining depreciated value of the article. The fixed rate is applied to the balance until the salvage value is reached and no further depreciation is possible; this method results in higher depreciation charges during the earlier life of the article and lower charges in later years and provides automatically for the salvage value at the end of the period.

Suppose annual depreciation is calculated by this method at the rate 20% for a tractor costing Rs. 24,000. We go on taking 20% of the remaining balance until salvage value is reached, as in this case about Rs. 4033/-.

4. Sum of the year digits method or Reducing Fraction Method

By this method the annual depreciation is found out by multiplying a fraction times the amount to be depreciated (cost minus the salvage value). We can determine the fraction for any year by the following formula:

Fraction for any year

$$= \frac{\text{The years of life remaining at the beginning of accounting period}}{\text{The sum of the years of the life of the asset}}$$

$$\text{Fraction for } 1^{st} \text{ year} = \frac{10}{1+2+3+4+5+6+7+8+7+9+10} = \frac{10}{55}$$

$$\text{Fraction for } 2^{nd} \text{ year} = \frac{9}{1+2+3+4+5+6+7+8+7+9+10} = \frac{9}{55}$$

Just as we compute depreciation by straight line method, similarly we deduct salvage value from original cost each year for finding the annual depreciation

Selection of the Depreciation Method

Selection of a particular deprecation method will depend upon two factors:

1. The method which best approximates the loss in the value of the asset;
2. Ease of computation.

Obviously the straight line method is the easiest to compute and understand. This is the most commonly used method. The diminishing balance method and reducing fraction method give similar results. The other methods of working out depreciation are compound interest method and depreciation according to use.

Some Special problems in Computation of Depreciation

1. The life of many machines is often longer than the normal expectations used in estimating the depreciation. Therefore, the farmers need to make suitable adjustments through time by judging the actual condition of the implement or the machine.
2. In case of minor repairs, use of oil and lubrication, etc., normally no extra depreciation is charged. A convenient procedure is to add the cost of the repairs to the existing value of the asset.
3. In case of small tools, they are considered to be used up within one year. Assume 100% depreciation on them.
4. If a machine is purchased second hand, the number of years it has been used and its original price may not be known, the logical procedure in this case is to estimate the remaining years of its life and begin depreciation on the purchase price.

5. When livestock is in a breeding stage or appreciation phase revaluation method is will be more useful. But when they start depreciating, say after 3 years in case of bullocks, straight line method will be more appropriate.
6. At the initial stage of appreciation, depreciation may not be charged because expenses for feeds, etc. are already being incurred. After attainment of the adult age depreciation must be charged.

Every method has its strong and weak points but for most of the purposes, normally straight line method is used.

Net Worth Statement

The net worth statement shows the financial condition and stability of the business at a particular point of time. Net worth is shown as an excess of assets over liabilities, i.e. the liability of the business to the farmer or the farmer's claim in the business. Net worth statement is, thus, prepared for the farmer and not for his business. It shows whether the business is expanding or shrinking. The business is said to be solvent when the net worth or equity is greater than zero. When the total liabilities are not covered by total resources, the business is said to be insolvent or bankrupt. The difference between assets and liabilities shows the distance from insolvency position; greater the difference on positive side, the sounder will be the business. The farmer is generally more concerned about the immediate solvency than the ultimate solvency. It is, therefore, necessary to classify the assets and liabilities to know the real position of the farm business at a point of time and compare the nature of the liabilities to the assets.

The assets and liabilities can be grouped as under :

Assets

a.) *Fixed assets :* Fixed assets are of the nature that it is difficult to convert them into cash to meet any current obligations. Examples: land, buildings, etc.

b) *Working assets :* These are more liquid than fixed assets, such as farm machinery, equipment, etc.,

c) *Current assets :* They are the most liquid assets and are consumable in a year. Examples: seeds, fertilizers, cash on hand, bills receivable, livestock for sale, etc.

Liabilities

a) ***Long duration liabilities :*** These liabilities are those which do not require repayment during the current the accounting period. Example is long-term loan.

b) ***Intermediate liabilities :*** These liabilities can be deferred or postponed for the present but fall due within the year. Examples are promissory notes and medium-term loans.

c) ***Current liabilities :*** Repayment of these liabilities may be demanded at once.

Thus, it will not be appropriate to buy a long lived asset having a slow turnover with funds obtained through floating bills or short time notes. The best policy would be to match the liabilities with the nature of assets and earnings from them in respect of time.

Analysis of Net Worth Statement : It is useful to find out ratios between different categories of assets and liabilities.

Net Capital Ratio : It is a ratio between total assets and total liabilities and it is worked out to measure the degree of financial safety over a period of time by comparing the present position of the business with that on some previous date.

The net capital ratios of two farmers are computed as under :

$$\textbf{Net Capital Ratio} = \frac{\text{Total assets}}{\text{Total liabilities}}$$

$$\text{Farmer 'A'} = \frac{\text{Rs.40,000}}{\text{Rs.30,000}} = 1.3$$

$$\text{Farmer 'B'} = \frac{\text{Rs.20,000}}{\text{Rs.10,000}} = 2.0$$

Although the net worth in both the cases is the same, yet the net capital ratios indicate that the business of farmer 'B' is in much safer position than that of farmer 'A'. For example, a decline of 25% in the prices of major assets would wipe out the margin of safety in case of the farmer 'A'. But in case of farmer 'B', it would take a decline of 50% in prices to place his business in an equally precarious position.

Working Ratio: It is derived by dividing the sum of working and current assets by the sum of medium-term liabilities and current assets. It measures the financial safety of the business over an intermediate period of time.

$$\textbf{Working Ratio} = \frac{\text{Sum of working and current assets}}{\text{Sum of medium-term liabilities and current liabilities}}$$

Current Ratio : It is obtained by dividing the current assets by the current liabilities. This ratio measures the degree of immediate solvency of the business.

$$\textbf{Current Ratio} = \frac{\text{Current assets}}{\text{Current liabilities}}$$

These ratios can also be compared over time. Wider these ratios are, the better will be the financial position of the business. The farmers should always try to improve the net worth of their business.

Two factors can improve the net worth of the farmer: i) changes in the general level of prices and ii) increase in investment through efficient production and by ploughing back the enhanced profits of the business.

Now farming is becoming more capital intensive requiring heavy investments in farm machinery. The farmers have to depend upon banks and other institutions for credit. For obtaining credit, the analysis of net worth statement is becoming more and more an important record. Bankers are interested in the "continuous net worth statement" in order to compare the financial stability and to estimate the management potential of the farmer.

Comparison of Balance Sheet and Income statement

S. No.	Balance sheet or Net worth statement	Income statement or Profit or loss statement
1.	Net worth statement is a summary of assets, liabilities and owner's equity at a given point of time.	Income statement is a summary of both cash and non cash financial transactions of farm business accrued during the selected accounting period
2.	The most commonly requested document by a lender in reviewing a loan request.	It is used to measure the financial profitability of business during a period of time.
3	It is used in preparation of income statement and tax returns	It is used in making an analysis of the business profitability, efficiency and financial stability.
4.	Net worth statement offers a little insight into financial transactions of accrued in business during an accounting period.	Information from this document is used in preparation of cash flow summary.

Cash flow statement

The cashflow statement is a measure of changes in cash the business has on hand from month to month. It records or projects all cash receipts less all cash disbursements. A business may use the cash flow statement as a record of what has occurred to cash or as a projection into the future to determine future needs for cash or as both.

The cashflow statement is accurate, when it is a record of past receipts and disbursements and an estimate, when it is projected for future months. The cashflow statement is usually calculated on a monthly basis for an entire year.

❑❑❑

Chapter 7

Forest Finance and Accounts Treatment and Classification of Special Transactions for Financial Accounts

Farm Financial Accounts or Farm Cash Accounts

These records relate to the annual operation of the business and the profits and losses associated with it. Whereas, the farm inventory is a record of the financial condition at a particular time, the farm cash account is a record of the results of the operation of the farm over a period of time, usually one season or one crop year.

Farm financial accounts can be as simple as just keeping a statement of the payments (expenses) and receipts. Detailed cash records will split up these expenses and receipts on the various enterprises as well as operations involved in the business.

Types of Cash Accounts : There can be many such records within the system of complete records for assembling facts needed in the operation of farms. Important records are as follows:

Farm Diary : Diary is meant for daily recording of all important events on the farm. It may include weather reports, any unusual events of the day, as well as cash receipts and expenses from all resources.

Farm Business Cash Record : It is a record of all cash receipts from and all cash expenses incurred on the farm business. The farm business cash record omits all personal cash expenses and all non-farm cash income. The objective of this record is to measure changes in the farm business only as far as cash transactions are concerned. The business cash record can be maintained in a simple from or a classified form.

Personal Cash Record : It is strictly a record of all personal expenses of the farm family, excluding all the farm costs and usually a recording of the cash income of the family from all sources, regardless of whether the income is from farm or nonfarm sources.

Complete Cash Record: It includes expenses on all items both personal and business and income from all sources recorded according to source types.

Use of Farm Financial Accounts

The following are the uses of farm cash accounts:

1. **Check on cash availability at any time :** The farm cash account provides as basis for checking the total cash on hand in the bank at any time. e.g. A complete cash account with one summary column for all personal expenses, can be used at the end any month as a running check on the cash receipts and cash expenses to-date to determine the amount of cash available for future operations.

2. **Provides receipts and expenses information for quick farm adjustments :** The farm cash account provides a complete record of receipts and expenses at any time.

   e.g. Sometimes, farm plans need to be changed immediately in order to adjust to changing physical and economic conditions. A cash account provides the facts needed to make these revisions on a sounder basis. After the consideration has been given to the expenses and receipts, it also supplies the facts needed to make an earlier and more accurate appraisal of changes that are contemplated. A frequent check on receipts compared with expenses will save time and money.

3. **Provide information on costs and returns of different enterprises:** The summarized cash account of the farm business will provide an indication of the relative importance of the receipts and expenses from different sources. The classified receipts compared with the classified expenses, will indicate major differences in enterprise costs and returns.

4. **Provide the basis for a simple study and analysis of farm business :** The summarized classified cash account of the farm business provides the basis for simple analysis of the farm business. The major objective of an analysis of any business is to provide a more intelligent basis for future planning.

A farm financial account is a brief, simple record of cash transactions involving the farm business as a whole and does not provide the detailed information needed for a study of all the different internal operations of the farm. Such a record, however, will supply much information of value in planning the future operations of the farm business.

5. The analysis provides farmer an opportunity to compare the years results with his expectations (i.e. his budget at the start of year).
6. It helps to make comparisons of his efficiency with performance of other farmers operating under similar conditions.
7. Analysis provides facts about the rates at which the farmer is combining resources into saleable products.

Treatment and Classification of Special Transactions for Financial Accounts

There are a number of transactions which are common to all farms and can be handled with standard procedures in farm financial accounting. **The treatment of certain transactions or items which cannot be handled in the financial accounts in the same manner as ordinary purchases or sales are discussed as below. Some of them are:**

1. Govt. payments like subsidies, etc.
2. Capital changes
3. Off-farm work and family labour
4. Joint farm and household expenses
5. Farm produce consumed in household
6. Non-farm income and expenses
7. Borrowed capital and interest payments
8. Prepayments and accruals (Income earned/profit)
9. Taxes
10. Losses

1. **Govt. payments like subsidies, etc. :** There are a large number of Govt. programmes and development schemes, through which

the farmers get cash payments in the form of prizes, premiums or certain concessions, such as land revenue remission. *Eg.* Premium on good quality seed, subsidy on purchase of improved seed, implements/equipments, minor irrigation loans.

2. **Capital changes :** Usually farm financial accounts include pages for recording business expenses and receipts for the year. They also include pages and forms for entering other costs, such as depreciation which do not show up as cash expenses during the year. e.g. Poultry farm (chicks) – feeding and selling (cots and returns) within a year.

3. **Off-farm work and family labour :** Most of the farmers and their family members have no full time work on their farms due to small size farm holdings. So, they have to depend on other incomes from off-farm work. This earning should be recorded separately for the purpose of analyzing the efficiency of the farm business.

4. **Joint farm and household expenses :** i.e. Radio, Newspaper, magazines, telephone, automobile, etc., are purchased mainly for the enjoyment of the family but they contribute market reports, technical information or other knowledge for family efficiency.

5. **Farm produce consumed in house hold :** Considerable number and quantities farm raised products are ordinarily consumed by the family, which are a part of the farm income as commodities (sold through regular market channels).

6. **Non-farm income and expenses :** Some farmers might have a small industry unit in addition to his farming business. So, transactions from that industry or enterprise should be kept separate.

7. **Borrowed capital and interest payment:** Interest is an expense to the farm business for the use of borrowed capital. It is a payment for the use of a resource owned by other persons and it is similar to wages for labour and rent for land. But borrowed capital is re-paid and it is not an expense.

8. **Pre-payments and accruals: Accruals :** Income may be earned during the year but not received until the next accounting period begins. If the true income of the business is to be determined, these credits must be entered in the accounts for the year the income accrues (other underestimated).

Advances or pre-payments must be recognized in the accounting period. e.g. A farmer gives some advance money to a hired man before closing account book and that should not be entered as debit against that current year (but noted).

9. **Taxes :** Income tax of net income payable has been computed. Taxes like crop cess, land revenue, etc. are also paid as farm expenses.
10. **Losses :** The common cost to farm business is the loss through death of animals and poultry birds, etc. The consumed feed during the year by the animals is recognized as expense.

Such transactions or happenings can be classified and accounted for by:

i) Mode of payment:	(a) If received in cash (subsidy, etc.), consider as receipt
	(b) If paid in the form of reduced costs, consider the reduced cost as expense
	(c) If the cost of any asset is taken as a subsidy, it may first be shown as receipt and then as expense – to show some value of the asset in the inventory
ii) Purpose of analysis	Treat the special transaction differently depending upon the purpose of analysis. e,g. whether for estimating net worth or for complete yearly business analysis.
iii) Capital Receipts	These are not included in the annual business analysis but taken care of in the inventory changes.

Farm Record Maintenance and Cost Analysis

Introduction

In context of farm management and natural resource economics, the farm record maintenance and cost analysis approach needs to be adopted with in conceptual framework of farm level agricultural system. Beyond traditional farm management; recently system analysis of farm management approach is widely accepted. Farm management is nothing new. It is something review every time we choose between alternatives as we attempt to make the most overlies, fully keeping consideration of natural resource. Farm management has shifted its focus from input-out relationships to enhancing an ecological balance and system resilience through such avenues as increasing diversity, component interactions, and enhanced biological sub-systems. Therefore, very right definition of farm management aims to ensure that the organization serves its mission (including environment) in an effective way, and also that it serve the needs of those who control or otherwise have power over the organization. Major features of farm management need consideration of maintaining farm records and accordingly make cost analysis. The most successful management is seen as installing both efficiency in resource use and effectiveness in the achievements of the goals. Also, management is seen as a universal process involving a set of management functions. These functions are considered in wide perspectives. As planning, organization, staffing, communication, motivation, leading, monitoring and controlling. The farm record maintenance and analysis is also useful in decisional roles at the different levels of stakeholders. A farm owner may use this information for direct supervision, and running of the organization. The important use of farm record maintenance and farm analysis are meaning but to know its relations with the various environment surrounding it, to develop the organization strategies which may be viewed as a mediating force between the organization and farmers and to carry out duties of supervision which they seek to tailor strategy to its strengths and to its need and trying to maintain a pace of change that is responsive to the environment without being disruptive to the organization.

Since farm management is considered a thought of as being a decision process in a continual ways, the maintaining farm records and cost analysis operated upon; will always help in decisions to be taken for allocating the limited resources of land, labour, and capital among alternatives and competitive uses. This allocating process forces the manager to identify goals (including environment) to guide direct the decision-making.

In the context of natural resources economics, the farm management has to be considered broadly. This is of farm management as the process by which resources and situations are manipulated over a time by the manager of the farm system with information recorded in farm record in ordered to achieve his/her goals. Thus, farm system looks the farm as having a variety of interface, i.e. Contact point with other systems or environment. The details on farm record maintenance and cost analysis are given on fourth part of chapter. Cost principles most of the producers give considerable importance to the cost of production while keeping production decision. When we speak of cost of producing a commodity, we generally refer to the expenses incurred in producing a unit of a product in a particular.

2. Time period, without specifying the amount and the time period, any reference to cost will be meaningless.

Cost consideration enters into almost every day in farm organization and operations. To get a clear picture of cost concepts that is appropriate to farm management decisions, the cost problems is divided into three parts:

i) Cost output relation.

ii) Cost price relation

iii) Product cost in joint farm operation and joint product.

Cost elements

Cost is made up of two elements:

i) Fixed cost and;

ii) Variable cost. the sum of these two costs is called total cost. The other costs are derived from these groups as listed below:

Table1: Hypothetical cost data showing in the relationship among the various cost concepts

Q units of output	Fixed cost FC	Variable cost VC	Total cost FC +VC	Average Cost per unit TC/Q	Margina l VC/Q	Average fixed cost FC/Q	Average variable cost VC/Q
0	125	0	125	Infinity	-	.	0
1	125	30	155	155	30	125	30
2	125	45	170	85	15	62.5	22.5
3	125	56	181	60.3	11	41.7	18.7
4	125	58	183	45.8	2	31.3	14.5
5	125	62	187	37.4	4	25.8	12.4
6	125	70	195	32.5	8	21.8	11.7
7	125	90	255	30.7	20	17.8	12.8
8	125	131	256	32.0	41	15.6	16.4
9	125	203	328	36.4	72	13.9	22.5
10	125	335	460	46.0	132	12.5	33.5

Table 1 information showed that :

i) Fixed cost is constant one all the levels of output.

ii) V.C. is associated with the level of output

iii) TC and VC increases with the increases in output.

iv) MC Is independent of FC, whatever may be the amount of the fixed cost.

v) AFC is continuously decreasing (i). Fixed cost (FC): sum of interest, depreciation, rent, traps, unpaid family labor, crop or live stock insurance, if any. (ii) variable cost (VC): sum of cost of seeds, manure, hired labor, irrigation changes and the like that are directly related to output. (iii). Total cost (TC) = F.C. + VC. Cost per unit of output: (iv). Average cost (AC) = TC/ Q (Q= Output) (v). Marginal cost (MC) VC/Q (vi). Average fixed cost (AFC) = FC/Q (vii). Average variable cost (AVC) =VC/Q

Fixed Cost (FC): These costs are related to fixed resources and are overhead cost. They are the same at all levels of production and result from costs that have already been sunk. Rent, interest, depreciation taxes and wages of the permanent laborers constitute fixed cost. Fixed costs have little relation to making decision on the level of production of farming practices.

Variable Cost (VC): These costs are related to the variable resources and change in the output. The variable costs are fit, if there is no production on the farm. They change with the quantity of production. In the beginning as the production increases, variable costs rise quite rapidly, but with further rise in production variable costs do not increase proper timely with the production due to economies drought about by mass production. Later on, as diminishing return set in variable cost start rising more rapidly than the production.

The variable costs are very important deciding what should produce & how much to produce. If farming is to be carried on, the variable cost must be less than the selling price.

Total Cost (TC): The fixed cost and variable costs make total cost of producing each unit of crop or livestock product. The total cost stands even when production is zero. They increase on the like variable costs & determine, if farming will be profitable. But once the total cost are covered, the farmers remains indifferent to the average cost as per unit cost of production.

Average Total Cost (ATC): It refers to the average of all costs (fixed plus variable) per unit of output. In the beginning the average costs are very high because the fixed costs are distributed on a few units of production. But as more units are produced the fixed cost are spread over more and more units. When the fixed costs have spread over many units, there is not much effect of the fixed cost on average costs. Variable cost assumes importance as average costs being to rise. Average cost finally began to rise as evident for the cost curves.

Average Fixed Cost (AFC) : Average fixed cost is a fixed cost per unit of output. Since the total fixed cost is the same at all the levels of production, the average fixed cost falls continually at a decreasing rate as more output is produced. It is because the fixed cost is divided by increasingly large numbers, as output increases.

Average Variable Cost (AVC) : The average variable cost refers to total variable cost per unit of output. The average variable cost has an inverse relationship with average Product (ap). When ap increases, avc decreases. When ap decreases, avc increases. Further more, when a.p. is at maximum, the avc must be at the minimum.

Marginal Cost (MC) : Marginal Cost (mc) is the change is cost associated with an increase of one unit of output. As marginal cost are related to the costs of producing additional units of output, they are affected only by the variable costs, fixed cost, as a rule, do not influences the marginal cost, because they neither increase nor decrease with the additional production. Marginal costs are very important in determining as to how far production should be pushed and how much of the various resources should be used. A farmer adds to production as long as the added return is greater or at least equal to the added cost.

Cost concept the cost concept approach to farm costly which has kindly been used by the directorate of economics & statistics and ca cp, ministry of agric. & costs, Govt. of India, is discussed below: cost a1 = all actual expenses in cash and kind includes in production by owner (a) wages of hired human labour. (b) wages of permanent labour. (c) wages of contract labour. (d) wages of hired bullock labour. (e) imputed value of owned bullock labour. (f) changes of hired machinery. (g) imputed value of owned machinery. (h) market rate of manure & fertilizers (i) market rate of seed. (j) imputed value of owned seeds. (k) imputed value of manures. (l) market value of pesticides, herbicides, hormones, etc. (m) irrigation charges. (n) irrigation structure, etc. (o) interest on working capital (p) miscellaneous expenses (value of other items which are used up in current production). Cost a2 = cost a1 + rent paid for leased in land is any. Cost b1 = cost a1 + interest on value of own capital (b1 = a1 + overhead cost) overhead cost includes = interest on fixed capital = interest on working capital = repairs to dead stock = depreciation cost b2 = cost b1 + rental value of owned land and rent paid for leased in land. Return to family labour or cost c1 = imputed value of family labour. Return to land or cost c2 = cost b2 + imputed value of family labour.

Break – Even Point Analysis

The break even point may be defined as the point, when sales revenue is equal to total cost (fixed and variable). In other words, it represents the level of activity, when there is neither any profit nor loss.

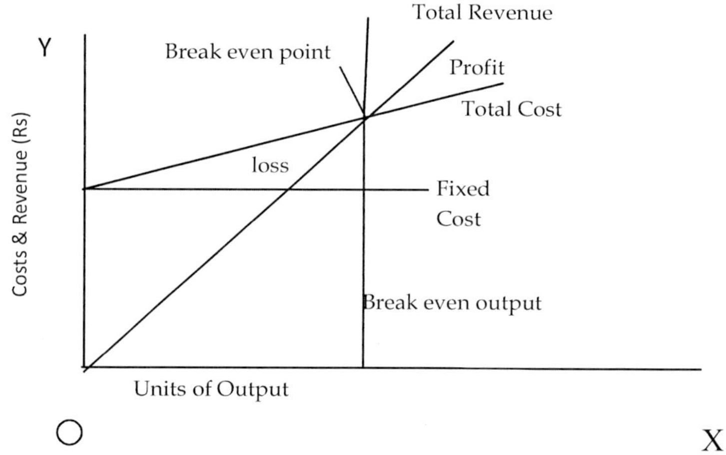

Fig. Break even point analysis

Selling Price	:	Rs. 200
Variable Cost	:	Rs. 300
ContributionPer unit	:	Rs. 100
F.C.	:	Rs. 1,50,000
B.E. Point	:	1,50,000 / 100
		1500 units

$$\text{Breakeven point} = \frac{\text{Fixed overhead}}{\text{Contribution}}$$

Contribution = Sales revenue – Variable cost

□□□

Chapter 8

Forest Efficiency Measures and Business Decision

Farm Efficiency Measures

An important demand in farm forest business management or decision-making relates to the manner in which available resources are allocated. A measuring indicator is necessary to provide the standard for apprising accuracy of decisions regarding the use of resources.

Efficiency can be related to :

1. The operation of farm business as a whole.
2. Any individual phase of the business, line of production or enterprise (e.g. Dairy, poultry, kokum juice, amla/tamarind pickles, etc.)
3. The use of various factors of production or resources (land labour capital)
4. Any single input (fertilizer, feeds, machine, etc.)

Various efficiency measures, therefore, need to be developed to express technical efficiency in various farms enterprises and relate these to the financial success.

The various farm efficiency measures can be grouped and discussed as:

1. Physical efficiency measures *(technical efficiency)*
2. Value efficiency measures *(financial efficiency)*

They can be further categorized into: a. Aggregate or absolute measures. b. Ratio measures.

1. Physical Efficiency Measures

a) Aggregate measures

i) Total area of the farm

ii) Number of livestock

iii) Total production

b) Ratio measures

i) **Land use efficiency**: Yield per acre, Production efficiency, Crop yield index, Intensity of cropping, Percentage of land under selected crops

ii) **Labour efficiency**: Crop acreage per man, Productive man-work equivalent

iii) **Machinery efficiency**: Horse power per acre of land available & used

2. Financial Efficiency Measures / Value Efficiency Measures:

a) Aggregate measures

i) Total capital managed

ii) Gross income

iii) Gross expenses

iv) Gross profit

v) Net worth = Total assets - Total liabilities

Farm Income and Profit Efficiency Measures

i) Net cash income = Total cash receipts - Total cash operating expenses

ii) Net farm income = Net cash income $\pm$ Change in inventory and depreciation

iii) Farm earning = Net farm income - Value of farm privileges

iv) Family labour earning = Farm earnings - Interest charges on farm capital

v) Returns to management = Family labour earnings - Imputed value of family labour

vi) Per cent return to capital = $\frac{\text{Farm earnings - Family labour value}}{\text{Average capital investment}} \times 100$

b) Ratio measures

i) Gross output per gross input

ii) Fertilizer cost per crop acres

iii) Power and equipment cost per crop acre

iv) Power and equipment / investment per acre

v) Cost Ratio

vi) Capital ration

vii) Income ratio

viii) Financial Solvency position

Evaluation of records and surveys for research

Farmers who maintain records fall in the category of higher income. Farm business data from the records facilitate management and contribute to the incomes. So, farm records are extremely helpful to the research workers and those records are usefully employed in the research on the following specific aspects :

i) **Area variation studies** to estimate the earnings and financial success of farming in different areas and under different types and systems of farming (i.e. **Relative profitability**)

ii) **Capital investment patterns studies** to determine capital investment needed for different types and sizes of farms and for different enterprises (i.e. **Farm financial management**).

iii) **Cost of production studies** to determine the cost of production per unit of various enterprises and cost of performing of various operations.

iv) **Economics of scale studies** to determine the effect of size, organization and productive efficiency on the earnings of the farmers.

v) **Resource requirement studies** to determine the physical input requirements in different enterprises, areas, different types and sizes of farms.

vi) **Historical trends** to determine the general direction of the farm business earnings (gaining or loosing)

vii) **Consumption pattern studies** to determine the family living costs in different areas.

❒❒❒

Chapter 9

Classification of Private Businesses, Legal form of Business Organizations. Stock and Stock-holders, Forms of Loans; Short and Long-Term Loans

Loan and its Classification

Meaning of Loan : Loan is the certain amount provided for certain purpose on certain condition with some interest and which should be repaired within a stipulated period.

Loan: *Money borrowed that is usually repaid with interest*

In finance, a **loan** is a debt provided by one entity (organization or individual) to another entity at an interest rate, and evidenced by a note which specifies, among other things, the principal amount, interest rate, and date of repayment. A loan entails the re-allocation of the subject asset(s) for a period of time, between the lender and the borrower.

In a loan, the borrower initially receives or *borrows* an amount of money, called the *principal,* from the lender, and is obligated to *pay back* or *repay* an equal amount of money to the lender at a later time. Typically, the money is paid back in regular *installments,* or partial repayments; in an annuity, each installment is the same amount.

The loan is generally provided at a cost, referred to as interest on the debt, which provides an incentive for the lender to engage in the loan. In a legal loan, each of these obligations and restrictions is enforced by contract, which can also place the borrower under additional restrictions known as loan covenants. Although this article focuses on monetary loans, in practice any material object might be lent.

Acting as a provider of loans is one of the principal tasks for financial institutions. For other institutions, issuing of debt contracts, such as bonds is a typical source of funding.

Classification based on Purpose

1. **Productive Loan :** This loan is supplied to increase the productivity of crop. These are also known as short term loans/ crop loan. These loans are repayable within period ranging from 6 to 18 months in lump sum.
2. **Marketing Loan :** These are mean for helping the farmers to overcome distress sale and market the produce in a better way.
3. **Consumption Loan :** Any loan advanced for purpose of other than production is known as consumption loan. e.g. Medical expenses.

Classification based on Time

1. **Short-Term Loan :** These loans are paid back within period ranging from 5 to 18 months. All crop loans are said to be short-term loan, these loans are usually repaid fully at one time in a current production year. This loan can be used for purchasing seeds, land protection measures, taxes, rent and other current farm expenses.
2. **Medium-Term Loan :** These loans are extended for a period 15 months to 5 years. These loan usually repaid in installment. It is needed for purchase of implement, electric motors, cattle, bunds and other land improvement repair of well, etc.
3. **Long-Term Loan :** In this credit period required is more than 5 years up to 15 to 20 years. This loan is meant for leveling of land designing of well, purchase of machineries, etc.

Classification based on Security

1. **Secured Loan:** Loan advanced against some security by the burrower is termed as secured loan.
2. **Unsecured Loan:** Based on confidence between the burrower and tender the loan transaction takes place.

Classification based on Liquidity

1. **Self Liquidity Loan :** The income generated through the loan help the farmer to repay the entire amount in the same year of obtaining loan.
2. **Partially Liquidating Loan :** The income generated through these burrower will help to plant of the loan component only.

Classification based on Activity Oriented

e. g. Dairy loan, Medicinal plant cultivation loan, Tractor loan, Sawmill loan, forest based cottage industry loan, etc.

Classification based on Approach

1. Individual Approach

   e. g. Crop loan
2. **Area Approach :** Here loans are advanced by taking area by branch. Service area followed by the bank is an apt
3. **DIR Loan :** Loans are advanced to the weaker sector of the community of an internal role of 4 % per annum.

Classification based on Contact with :

1. Direct Loan
2. Indirect Loan

Other types of Loan

Secured loan : A secured loan is a loan in which the borrower pledges some asset (e.g. a car or property) as collateral.

Mortgage loan : Mortgage loan is a very common type of debt instrument, used by many individuals to purchase housing. In this arrangement, the money is used to purchase the property. The financial institution, however, is given security—a lien on the title to the house—until the mortgage is paid off in full. If the borrower defaults on the loan, the bank would have the legal right to repossess the house and sell it, to recover sums owing to it.

In some instances, a loan taken out to purchase a new or used car may be secured by the car, in much the same way as a mortgage is secured by housing. The duration of the loan period is considerably shorter—often corresponding to the useful life of the car. There are two types of auto loans, direct and indirect. A direct auto loan is where a bank gives the loan directly to a consumer. An indirect auto loan is where a car dealership acts as an intermediary between the bank or financial institution and the consumer.

Unsecured loans : Unsecured loans are monetary loans that are not secured against the borrower's assets. These may be available from financial institutions under many different guises or marketing packages:

- Credit card debt
- Personal loans
- Bank overdrafts
- Credit facilities or lines of credit
- Corporate bonds (may be secured or unsecured)

The interest rates applicable to these different forms may vary depending on the lender and the borrower. These may or may not be regulated by law. In the United Kingdom, when applied to individuals, these may come under the Consumer Credit Act, 1974.

Interest rates on unsecured loans are nearly always higher than for secured loans, because an unsecured lender's options for recourse against the borrower in the event of default are severely limited. An unsecured lender must sue the borrower, obtain a money judgement for breach of contract, and then pursue execution of the judgement against the borrower's unencumbered assets (that is, the ones not already pledged to secured lenders). In insolvency proceedings, secured lenders traditionally have priority over unsecured lenders when a court divides up the borrower's assets. Thus, a higher interest rate reflects the additional risk that in the event of insolvency, the debt may be uncollectible.

Demand loans : Demand loans are short-term loans that are atypical in that they do not have fixed dates for repayment and carry a floating interest rate which varies according to the prime lending rate. They can be "called" for repayment by the lending institution at any time. Demand loans may be unsecured or secured.

Subsidized loan : A subsidized loan is a loan on which the interest is reduced by an explicit or hidden subsidy. In the context of college loans in the United States, it refers to a loan on which no interest is accrued while a student remains enrolled in education.

Concessional loan : A concessional loan, sometimes called a "soft loan," is granted on terms substantially more generous than market loans either through below-market interest rates, by grace periods or a combination of both. Such loans may be made by foreign governments to poor countries or may be offered to employees of lending institutions as an employee benefit.

Equated Monthly Installment (EMI)

Equated Monthly Installment (EMI) is the amount paid by borrowers, each month to lender of the loan. EMI can be calculated for a car loan, home loan, personal loan, gold loan, etc. The principal amount borrowed and the interest are added and then divided by the number of months, an individual desires to pay the total amount. The installments thus derived are called EMIs.

How is it calculated?

EMI comprises of two variable components those are principal amount and interest rate. The component of interest amount is higher in initial years and decreases over the years. EMI is calculated using the factors like interest rates, loan amount and the tenure of the repayment.

Forms of the Business Organizations

There are three basic forms of business organization methods: the sole proprietorship, the partnership, and the corporation. With only a few limited exceptions, any type of business venture can use any form of business organization.

The factors that will affect the business form chosen are:

1. Ease of formation
2. Exposure to financial risk
3. Ability to raise capital
4. Tax treatment of income
5. Continuity of business upon the death of owner.

Types of Farm Business Organizations

1. ***Sole Proprietorship*** : A one-person operation. The business may have a number of employees or hired persons; but the proprietor owns, runs, and manages the business.
2. ***Partnership*** : An aggregation of owners. Two or more persons contribute their assets to the business and may share the management, responsibility, profits, and losses. Each partner pledges faith in the other partners and stands liable for the actions of all partners within the scope of partnership activities.
3. ***Limited Partnership*** : A special form of partnership permitted by state law to have one or more partners whose liability for partnership debts and obligations is limited to their investment in the business. A limited partner is just an investor. If a limited partner participates in management, then liability will exist for all partnership obligations like a general partner. A limited partnership must have at least one general partner who handles the management of the business and who is fully liable for all partnership debts and obligations.
4. ***Corporation*** : An artificial being created under state law. A corporation is a separate business entity distinct from its owners, who are called shareholders because they own shares or interests in the corporation. The major characteristic of the corporate form of business organization is this sharp line of distinction between the business and the owners. The corporation is a separate legal entity as well as a separate tax-payer.
5. ***Tax-Option Corporation*** : A creation of federal tax law. A corporation in all aspects except that the corporate entity pays no income tax because each shareholder owner reports his or her share of corporate income for income tax purposes on their individual income tax returns. Tax-option corporations are subchapter S corporations.

LEGAL Forms of Business Organization

The Four Common Forms of Business Organizations

1. The Sole Proprietorship

a) No legal formalities are required to form a sole proprietorship, though a fictitious name may need to be registered.

b) The sole proprietorship may only have one owner, but there may be many employees.

c) The personal assets of the owner are subject to claims against the business.

2. The General Partnership

a) A partnership is a "a voluntary association of two or more persons to carry on a business for profit"; the two types of partnership are:

   1. A general partnership where all the partners are general partners; and

   2. A limited partnership, which has at least one general partner and at least one limited partner.

b) No legal formalities are required to form a general partnership; however, a fictitious name statute may require the firm name to be registered to prevent confusion and/or fraud.

c) There is often a written general partnership agreement called the "articles of partnership", but they are not mandatory.

d) The Uniform Partnership Act (UPA) specifies the terms which govern the relationship of the general partners, where they have no agreement or have not agreed upon or anticipated a particular situation; the UPA has been adopted in 48 states and the District of Columbia.

e) Partnerships are not separate legal entities for most purposed nor do they pay taxes; they do file "information returns" with the IRS to report each partner's share of profits and thereby establish the amount of income from the partnership which the partner must declare as regular income.

f) Each general partner is an agent for the firm and other partners for partnership affairs.

g) The personal assets of the general partners are subject to claims against the partnership.

3. Limited Partnership

a) The general partner (s) manage the firm and have unlimited personal liability for claims against the business.

b) Limited partners do not participate in the management and do not have personal liability for claims against the business (of course a limited partner will lose his/her limited liability, if he/she participates regularly in significant managerial decisions.

c) Uniform Limited Partnership Act (ULPA) was widely adopted by the states as has been its 1976 Revision (RULPA).

4. The Corporation

a) Corporations are treated as being separate legal persons apart from their owners.

b) Corporations are chartered under state law.

c) A Subchapter S Corporation is a corporation which is not separately taxed from its owners (i.e. double taxation avoided) ; however, it is only available to closely held corporations with a small number of shareholders (75 or less).

d) For good cause, courts can pierce the corporate veil, disregarding the corporate entity; this may happen when a court determines:

e) Parent corporations have limited liability for claims against subsidiary corporations but the corporate veil can be pierced where a court determines, it is appropriate to do so.

f) Miscellaneous Corporate attributes.

g) Reform of Corporations: Strengthening the Board: Some corporate reform proposals focus on limiting the ability of management to engage in wrongful acts by giving a greater role in corporate governance to outsiders and/or creating a more independent board of directors which could exercise its power to restrict management.

Types of privately owned business

1. **Sole proprietorship:** A sole proprietorship is a business owned by one person. The owner may operate on his or her own or may employ others. The owner of the business has total and unlimited personal liability of the debts incurred by the business. This form is usually relegated to small businesses.
2. **Partnership :** A partnership is a form of business in which two or more people operate for the common goal of making profit.

Each partner has total and unlimited personal liability of the debts incurred by the partnership. There are three typical different types of classifications for partnerships: general partnerships, limited partnerships, and limited liability partnerships.

3. **Corporation :** A business corporation is a for-profit, limited liability or unlimited liability entity that has a separate legal personality from its members. A corporation is owned by multiple shareholders and is overseen by a board of directors, which hires the business's managerial staff. Corporate models have also been applied to the state sector in the form of Government-owned corporations. A corporation may be privately held (that is, close - that is, held by a few people) or publicly traded.

A comparison of the three forms of Business organization :

Factors	**Sole Propreitorship**	**Partnership**	**Corporation**
Number of owners	One	Two or more	One or more
Ease of formation	Very easy	Easy to very easy	Moderate to easy
Degree of financial risk	Owner is responsible for all debts	Partners share responsibility for debts	Owners have only limited responsibility for debts in name of company
Ability to borrow funds	Limited ability of owner to borrow	Improved ability to borrow due to multiple owners	Can borrow funds directly from public through shares and debentures
Legal requirements	None, except for possible declaration of name	Same as sole proprietorship; written contract between partners recommended	Charter must be secured from state
How taxes are handled	As part of owner's personal tax return	Profits are divided up among owners and are taxed on each partner's tax return	Pays own income tax
Continuity of business	Business is terminated at death of owner	Business may be terminated at death of any partner, unless provisions for continuation are made	Business continues after death of any stockholder

Stock and Stock-holders

A **shareholder** or **stockholder** is an individual or institution (including a corporation) that legally owns a share of stock in a public or private corporation. **Shareholders** are the owners of a limited company. They buy shares which represent part ownership of a company. Stockholders are granted special privileges depending on the class of stock.

These rights may include :

- The right to sell their shares.
- The right to vote on the directors nominated by the board.
- The right to nominate directors (although this is very difficult in practice because of minority protections) and propose shareholder resolutions.
- The right to dividends, if they are declared.
- The right to purchase new shares issued by the company.
- The right to what assets remain after a liquidation.

Stockholders or shareholders are considered by some to be a subset of stakeholders, which may include anyone who has a direct or indirect interest in the business entity. For example, employees, suppliers, customers, the community, etc., are typically considered stakeholders because they contribute value and/or are impacted by the corporation.

Shareholders in the primary market who buy IPOs (Initial Public Offering) provide capital to corporations; however, the vast majority of shareholders are in the secondary market and provide no capital directly to the corporation.

Stock market

A **stock market** or **equity market** is the aggregation of buyers and sellers (a loose network of economic transactions, not a physical facility or discrete entity) of stocks (shares); these are securities listed on a stock exchange as well as those only traded privately.

Size of market

Stocks are partitioned in various ways. One common way is by the country where the company is domiciled. For example, Nestle,

Roche, and Novartis are domiciled in Switzerland, so they are part of the Swiss stock market.

The size of the world stock market was estimated at about $36.6 trillion at the beginning of October 2008. The *total* world derivatives market has been estimated at about $791 trillion face or nominal value, 11 times the size of the entire world economy. The value of the derivatives market, because it is stated in terms of *national values*, cannot be directly compared to a stock or a fixed income security, which traditionally refers to an actual value. Moreover, the vast majority of derivatives 'cancel' each other out (i.e., a derivative 'bet' on an event occurring is offset by a comparable derivative 'bet' on the event *not* occurring). Many such relatively illiquid securities are valued as marked to model, rather than an actual market price.

Stock exchanges

Stocks are listed and traded on stock exchanges which are entities of a corporation or mutual organization specialized in the business of bringing buyers and sellers of the organizations to a listing of stocks and securities together.

Generally considered major stock exchanges are the Amsterdam Stock Exchange, **London Stock Exchange, New York Stock Exchange, Paris Bourse, Bursa Malaysia** (formerly Kuala Lumpur Stock Exchange) and **the Deutsche Börse** (Frankfurt Stock Exchange) and **Toronto Stock Exchange**. In Africa, examples include Nigerian Stock Exchange, JSE Limited, etc. Asian examples include the Philippine Stock Exchange, the Singapore Exchange, the Tokyo Stock Exchange, the Hong Kong Stock Exchange, the Shanghai Stock Exchange, and the Bombay Stock Exchange. In Latin America, there are such exchanges as the BM&F Bovespa and the BMV. Australia has a national stock exchange, the Australian Securities Exchange, due to the size of its population. The stock exchange is based in Sydney.

Market participants include individual retail investors and traders, institutional investors, such as mutual funds, banks, insurance companies and hedge funds, and also publicly traded corporations trading in their own shares. Some studies have suggested that institutional investors and corporations trading in their own shares generally receive higher risk-adjusted returns than retail investors.[4]

Trade

Trade in Market Simply means a price in which both parties (buyer and seller) agrees to deal for a product, in this case, the transaction of a share. Participants in the stock market range from small individual stock investors to large hedge fund traders, who can be based anywhere in the world. Their orders usually end up with a professional at a stock exchange, who executes the order of buying or selling.

Some exchanges are physical locations where transactions are carried out on a trading floor, by a method known as open outcry. This type of auction is used in stock exchanges and commodity exchanges where traders may enter "verbal" bids and offers simultaneously. An example of such an exchange is the New York Stock Exchange. The other type of stock exchange is a virtual kind, composed of a network of computers where trades are made electronically via traders. An example of such an exchange is the **National Association of Securities Dealers Automated Quotation System** (NASDAQ). **NASDAQ is an American stock exchange. It is the largest electronic screen-based equity securities trading market in the United States. With approximately 3,200 companies, it lists more companies and has more trading volume per day than any other stock exchange in the world. It was founded in 1971, by the National Association of Securities Dealers (NASD).**

Actual trades are based on an auction market model, where a potential buyer *bids* a specific price for a stock and a potential seller *asks* a specific price for the stock. Buying or selling *at market* means you will accept *any* ask price or bid price for the stock, respectively. When the bid and ask prices match, a sale takes place, on a first-come-first-served basis, if there are multiple bidders or askers at a given price.

The purpose of a stock exchange is to facilitate the exchange of securities between buyers and sellers, thus providing a marketplace (virtual or real). The exchanges provide real-time trading information on the listed securities, facilitating price discovery.

The New York Stock Exchange (NYSE) is a physical exchange, with a hybrid market for placing orders both electronically and manually on the trading floor. Orders executed on the trading floor enter by way of exchange members and flow down to a floor broker, who goes to the floor trading post specialist for that stock to trade the order. The specialist's job is to match buy and sell orders using open outcry. If a spread exists, no trade immediately takes place—in this case the specialist should use his/her own resources (money or stock) to close the difference after his/her judged time. Once a trade has been made the details are reported on the "tape" and sent back to the brokerage firm, which then notifies the investor who placed the order. Although there is a significant amount of human contact in this process, computers play an important role, especially for so-called "program trading".

The NASDAQ is a virtual listed exchange, where all of the trading is done over a computer network. The process is similar to the New York Stock Exchange. However, buyers and sellers are electronically matched. One or more NASDAQ market makers will always provide a bid and ask price at which they will always purchase or sell 'their' stock.[5]

The Paris Bourse, now part of Euronext, is an order-driven, electronic stock exchange. It was automated in the late 1980s. Prior to the 1980s, it consisted of an open outcry exchange. Stockbrokers met on the trading floor or the Palais Brongniart. In 1986, the CATS trading system was introduced, and the order matching process was fully automated.

From time to time, active trading (especially in large blocks of securities) have moved away from the 'active' exchanges. Securities firms, led by UBS AG, Goldman Sachs Group Inc. and Credit Suisse Group, already steer 12 per cent of U.S. security trades away from the exchanges to their internal systems. That share probably will increase to 18 per cent by 2010 as more investment banks bypass the NYSE and NASDAQ and pair buyers and sellers of securities themselves, according to data compiled by Boston-based Aite Group LLC, a brokerage-industry consultant.

Now that computers have eliminated the need for trading floors like the Big Board's, the balance of power in equity markets is shifting. By bringing more orders in-house, where clients can move big blocks of stock *anonymously*, brokers pay the exchanges less in fees and capture a bigger share of the $11 billion a year that institutional investors pay in trading commissions.

Market participants

Market participants include individual retail investors, institutional investors, such as mutual funds, banks, insurance companies and hedge funds, and also publicly traded corporations trading in their own shares. Some studies have suggested that institutional investors and corporations trading in their own shares generally receive higher risk-adjusted returns than retail investors.

A few decades ago, worldwide, buyers and sellers were individual investors, such as wealthy businessmen, usually with long family histories to particular corporations. Over time, markets have become more "institutionalized"; buyers and sellers are largely institutions (e.g., pension funds, insurance companies, mutual funds, index funds, exchange-traded funds, hedge funds, investor groups, banks and various other financial institutions).

The rise of the institutional investor has brought with it some improvements in market operations. There has been a gradual tendency for "fixed" (and exorbitant) fees being reduced for all investors, partly from falling administration costs but also assisted by large institutions challenging brokers' oligopolistic approach to setting standardized fees.

The Bombay Stock Exchange is Asia's first stock exchange established in 1875

In 12th century, France, the *courretiers de change* were concerned with managing and regulating the debts of agricultural communities on behalf of the banks. Because these men also traded with debts, they could be called the first brokers. A common misbelief is that in late 13th century, Bruges commodity traders gathered inside the house of a man called *Van der Beurze*, and in 1409, they became the "Brugse Beurse", institutionalizing what had been, until then, an informal meeting, but actually, the family Van der Beurze had a building in Antwerp where those gatherings occurred;[7] the Van der Beurze had Antwerp, as most of the merchants of that period, as their primary place for trading. The idea quickly spread around Flanders and neighbouring counties and "Beurzen" soon opened in Ghent and Rotterdam.

In the middle of the 13th century, Venetian bankers began to trade in government securities. In 1351, the Venetian government outlawed spreading rumors intended to lower the price of government funds. Bankers in Pisa, Verona, Genoa and Florence also began trading in government securities during the 14th century. This was only possible because these were independent city states not ruled by a duke but a council of influential citizens. Italian companies were also the first to issue shares. Companies in England and the Low Countries followed in the 16th century.

The Dutch East India Company (founded in 1602) was the first joint-stock company to get a fixed capital stock and as a result, continuous trade in company stock occurred on the Amsterdam Exchange. Soon thereafter, a lively trade in various derivatives, among which options and repos, emerged on the Amsterdam market. Dutch traders also pioneered short selling a practice which was banned by the Dutch authorities as early as 1610.[8]

There are now stock markets in virtually every developed and most developing economies, with the world's largest markets being in the United States, United Kingdom, Japan, India, China, Canada, Germany (Frankfurt Stock Exchange), France, South Korea and the Netherlands.

Importance of stock market

The main trading room of the Tokyo Stock Exchange, where trading is currently completed through computers. The **stock market** is one of the most important sources for companies to raise money. This allows businesses to be publicly traded, or raise additional financial capital for expansion by selling shares of ownership of the company in a public market. The liquidity that an exchange affords the investors gives them the ability to quickly and easily sell securities. This is an attractive feature of investing in stocks, compared to other less liquid investments. Some companies actively increase liquidity by trading in their own shares.

History has shown that the price of shares and other assets is an important part of the dynamics of economic activity, and can influence or be an indicator of social mood. An economy where the stock market is on the rise is considered to be an up-and-coming economy. In fact, the stock market is often considered the primary indicator of a country's economic strength and development.

Rising share prices, for instance, tend to be associated with increased business investment and vice versa. Share prices also affect the wealth of households and their consumption. Therefore, central banks tend to keep an eye on the control and behaviour of the stock market and, in general, on the smooth operation of financial system functions. Financial stability is the raison d'être of central banks. Exchanges also act as the clearing house for each transaction, meaning that they collect and deliver the shares, and guarantee payment to the seller of a security. This eliminates the risk to an individual buyer or seller that the counterparty could default on the transaction.

The smooth functioning of all these activities facilitates economic growth in that lower costs and enterprise risks promote the production of goods and services as well as possibly employment. In this way the financial system is assumed to contribute to increased prosperity.

Relation of the stock market to the modern financial system

The financial system in most western countries has undergone a remarkable transformation. One feature of this development is disintermediation. A portion of the funds involved in saving and financing, flows directly to the financial markets instead of being routed via the traditional bank lending and deposit operations. The general public interest in investing in the stock market, either directly or through mutual funds, has been an important component of this process.

Statistics show that in recent decades, shares have made up an increasingly large proportion of households' financial assets in many countries. In the 1970s, in Sweden, deposit accounts and other very liquid assets with little risk made up almost 60 per cent of households' financial wealth, compared to less than 20 per cent in the 2000s. The major part of this adjustment is that financial portfolios have gone directly to shares but a good deal now takes the form of various kinds of institutional investment for groups of individuals, e.g., pension funds, mutual funds, hedge funds, insurance investment of premiums, etc.

The trend towards forms of saving with a higher risk has been accentuated by new rules for most funds and insurance, permitting a higher proportion of shares to bonds. Similar tendencies are to be found in other industrialized countries. In all developed economic systems, such as the European Union, the United States, Japan and other developed nations, the trend has been the same: saving has moved away from traditional (government insured) "bank deposits to more risky securities of one sort or another".

I. The Economic Functions of the Stock Market

A. The stock market allows nearly anyone to participate in the risks and opportunities of corporate America. Real returns for the past two centuries have averaged 7 per cent per year.

B. How the Stock Market Works for Savers and Investors?

1. Savers invest in the stock market as a strategy to build wealth.
2. Investors can buy a diverse portfolio of shares and hold them over long periods of time, substantially reducing risk.
3. Small investors can purchase stock in an equity mutual fund, a corporation that buys and holds shares of stock in many firms.
4. Equity mutual funds have reduced the risk of stock ownership and attracted large amounts of funds into the market, helping to push stock prices upward.

C. How the Stock Market Works for Corporations?

1. To raise money, a corporation can a) use retained earnings, b) borrow money, or c) sell stock. Buyers can, if they wish, later resell shares on the stock market.
2. Each share of the stock is a fractional share in the firm's future net revenues.
3. People buy the stock of a corporation to get future dividends paid from corporate earnings and gains derived from increases in share prices.
4. The decisions of a firm's executives influence the firm's stock price. When investors (and their advisors and fund managers) believe that the decisions of corporate managers will increase the firm's future income, they will buy more of the stock, driving its price up. When investors believe that bad decisions are being made, the reverse happens and the stock's price falls.
5. Corporate board members are usually stockholders, and top managers are often given stock options. The value of the stock

options will rise sharply as the firm's stock price increases. This helps bring the interest of corporate decision makers into harmony with other stockholders.

D. How the Stock Market Works for the Economy?

1. The stock market benefits stockholders, helps discipline corporate decision-makers to be more efficient, and to undertake productive projects.
2. The price of a corporation's shares constantly sends signals to the listed corporation's board of directors and managers. Changing stock prices reward good decisions and penalize bad ones.
3. To increase the firm's value, the firm must undertake productive projects.

II. Stock Prices

a) Underlying the price of a firm's stock is the present value of the firm's expected future net earnings, or profit.

b) The value of a share depends on (1) the expected size of future net earnings, (2) when these earnings will be achieved, and (3) the interest rate by which the investor discounts the future income.

c) The PE (price/earnings) ratio of a firm is a general measure of the price of the stock compared to the actual asset value of the company. A PE below the sector average represents an 'undervalued' stock; a PE ration above the sector represents an 'overvalued' stock.

d) The PEG (PE/Growth) ratio is a general measure of the price of the stock versus its future expected earnings. The closer to 1 the PEG ratio is, the more the stock price represents actual predicted future value.

III. Stock Analysis

a) Stockholder who purchase stock for long-term gain from dividends and capital gains are investors; stockholder who purchase stock for short-term gain from capital gains are speculators.

b) Speculators are primarily concerned with the value and volatility of the stock itself; investors are more concerned with the health and growth of the company the stock represents.

c) Speculators use Technical analysis of the quantitative aspects of the stock's price over time to go long or sell short.

d) Investors use Fundamental analysis of both the historical quantitative and qualitative aspects of a company to buy shares in a company.

- Quantitative analysis looks at the historical trends of the company's stock in terms of prices and dividends
- Qualitative analysis looks at the business, management, and philosophy of the company as a whole to predict future trends.
- Value stocks are held by investors for the long-term accumulation of consistent dividends; growth stocks are held by investors in anticipation of long-term capital gains from increase in the price of the company's stock.

❑❑❑

Chapter 10

Co-operative Businesses, Time Element in Business, Business Risks

A co-operative is a business organization owned by the members who use the services of the co-operative. Control rests equally with all members and surplus earnings are shared by members in proportion to the degree they use the services.

Basic Principles of A Co-operative

All co-operatives are guided by the same general principles as stated by the International Co-operative Alliance. These principles are:

Voluntary and Open Membership – Co-operatives are voluntary organizations, open to all persons able to use their services and willing to accept the responsibilities of membership.

Democratic Member Control – Co-operatives are democratic organizations controlled by their members, who actively participate in setting their policies and making decisions.

Member Economic Participation – Members contribute equitably to, and democratically control, the capital of their co-operative. At least part of that capital is usually the common property of the co-operative.

Autonomy and Independence – Co-operatives are autonomous organizations controlled by their members. If they enter into agreements with other organizations, including governments, or raise capital from external sources, they do so on terms that ensure democratic control by their members and maintain their co-operative autonomy.

Education, Training and Information – Co-operatives provide education and training for their members, elected representatives, managers and employees so they can contribute effectively to the development of their co-operatives.

Co-operation Among Co-operatives – Co-operatives serve their members most effectively, and strengthen the co-operative movement by working together through local, national, regional, and international structures.

Concern for Community – Co-operatives work for the sustainable development of their communities through policies approved by their members.

Types of Co-operatives

Co-ops exist in almost every sector of the economy. The various types of co-operatives are listed below:

Profit Verses Non-Profit Co-operatives: Co-operatives can be formed as either for profit or not for profit co-operatives. In both cases the goal of the co-operative is to operate profitably, however, the surplus revenues are handled differently in these two types of co-operatives.

For-Profit Co-operatives: surpluses of the enterprise, after payment of dividends on shares and interest on loans, are distributed at the discretion of the directors among the members in proportion to their business transactions with the co-operative during the fiscal year.

Not-For-Profit Co-operatives: operating surpluses of the enterprise are not distributed to the members and must be returned to the co-operative's general reserve. The co-op can use these funds to improve services to members, distribute them to other non-profit co-operatives or give them to charity.

Types of Co-operatives

Producer Co-operatives

1. Some co-operatives process and/or market, their members' products and services directly.
2. Others may also sell the input necessary to their members' economic activities (such as agricultural supply co-operatives).
3. Owned by members who either purchase their inputs or supply the products and services they market.

Examples: Agricultural processing and marketing, farm supply, crafts, livestock and crops, seed cleaning, pooling of equipment.

Service Co-operatives

They provide services to their members (individuals or corporations) :

a) They are owned by the users of the services.

b) They can range in size from small day care centres to large housing co-operatives.

Examples : Community development, electricity, natural gas housing, day care centres and nurseries, health care, transportation and communication, tourism.

Financial Co-operatives

1. These co-operatives offer financial, loan or investment services, and insurance services to their members.
2. They are owned by user members or by subscribers to insurance.

Examples: Credit unions, insurance co-operatives, mutual companies.

Multi-Stakeholder Co-operatives

1. The membership of these co-operatives is made of different categories of members who share a common interest in the organization; i.e. clients, workers, investors, community organizations, etc.

Examples: Home care services, health services, community services

Consumer Co-operatives

a) They provide their members with goods for their personal use.

b) They are owned by the consumers of the goods sold by the co-operative.

Examples: Food, natural food, school supplies (stationery, school supplies, computers and software), hardware, clothing.

Worker Co-operatives

a) The purpose of these co-operatives is to provide their members with work by operating an enterprise. They may be found in all economic sectors.

b) These co-operatives are owned by their employee members. They are mainly part of the small or medium-sized enterprise sector.

Examples : Agri-food, natural food, manufacture and sale of clothing, communications and marketing, industrial production and manufacturing, home nursing services

Organizational Structure

A co-operative has 4 key groups of people:

Members: Members, as owner-users, are the reason the co-op is organized. They justify its continued existence through their patronage, capital investment, and participation in decision making. The members are represented in the co-op by the Board of Directors.

The Board of Directors: The Board of Directors is a policy-making body elected by the members, who represent the members by overseeing the co-op's business affairs. Usually 7–9 members (minimum 3) are elected to provide leadership. Directors, as trustees, establish policy, report to members and give direction to the co-op's hired management, generally without getting involved with the day-to-day operations.

Members of the board are held accountable for their actions by provincial and federal laws applying to businesses and by the by-laws of their co-op. For larger co-ops, the board of directors also retains an independent auditor to evaluate the co-op's financial condition. They may also appoint committees for such continuing concerns as property, finance or member relations, education or appoint ad hoc committees for fund raising or special projects. The board is directly responsible for hiring the manager but delegates the responsibility for choosing co-op staff to the management.

Management: Management supervises and co-ordinates the day-to-day operations of the co-op and are supervised by the board of directors.

❑❑❑

Chapter 11

Forest Taxation Tariffs and Insurance

Tax: A **tax** (from the Latin *taxo*) is a financial charge or other levy imposed upon a tax-payer (an individual or legal entity) by a state or the functional equivalent of a state to fund various public expenditures. A failure to pay, or evasion of or resistance to taxation, is usually punishable by law. Taxes are also imposed by many administrative divisions. Taxes consist of direct or indirect taxes and may be paid in money or as its labour equivalent. A few countries impose almost no taxation at all, such as the United Arab Emirates and Saudi Arabia.

There are two types of taxes in India :

1. **Direct tax**
2. **Indirect tax**

1. **Direct tax:** These taxes are directly charged on people and firms.

   a) **Personal income tax** : Personal income tax is often collected on a pay-as-you-earn basis, with small corrections made soon after the end of the tax year. These corrections take one of two forms: payments to the government, for tax-payers who have not paid enough during the tax year; and tax refunds from the government for those who have overpaid. Income tax systems will often have deductions available that lessen the total tax liability by reducing total taxable income. They may allow losses from one type of income to be counted against another. For example, a loss on the stock market may be deducted against taxes paid on wages. Other tax systems may isolate the loss, such that business losses can only be deducted against business tax by carrying forward the loss to later tax years.

   b) **Corporate tax**: Corporate tax refers to income, capital, net worth, or other taxes imposed on corporations. Rates of tax and the taxable base for corporations may differ from those for individuals or other taxable persons.

c) **Wealth tax :** Some countries' governments will require declaration of the tax payers' balance sheet (assets and liabilities), and from that exact a tax on net worth (assets minus liabilities), as a percentage of the net worth, or a percentage of the net worth exceeding a certain level. The tax may be levied on "natural" or legal "persons".

2. **Indirect tax :** These are the taxes that are charged on goods and services i.e. charged indirectly on people and firms. Indirect taxes are the main sources of tax revenue for Government of India. The following are the indirect taxes :

   i) **Excise duty**: Taxes are charged on the production of goods

   ii) **Sales tax:** Taxes are imposed on goods at the time of sale.

   iii) **Service tax:** These taxes are imposed on services.

   iv) **Property tax:** Taxes are imposed on real estate.

   v) **Customs duty:** These taxes are imposed on the import of goods.

A brief guide to forestry taxation

Income Tax : Income realized through the sale of timber is exempt from income tax. No income tax relief is available against capital purchase or annual expenditure costs. Payments received under forestry grant schemes are tax free, with the exception of payments made in compensation for agricultural income foregone and revenue from other sources, such as rents which are taxable.

Inheritance Tax (IHT) : Following two years of ownership, commercial woodlands in individual ownership will normally attract 100% IHT Business Property Relief, thus removing them completely from any charge to IHT. If death occurs within the first two years of ownership, the IHT liability can ordinarily be paid over ten years in interest free installments.

Capital Gains Tax (CGT) : There is no CGT liability on the gain in the value of commercial tree crops. Any gain realized on the disposal of woodland is split between:

- Gain attributable to trees - on which no CGT is payable
- Gain attributable to the land - on which CGT is levied

Where funds realized on the sale of a qualifying business asset are re-invested in a new qualifying business asset upto twelve months before or three years after the sale, then any chargeable gain from that sale can in principle be deferred until such time as the second asset is sold.

Forestry is considered as an eligible trade for which this roll-over relief can be applied, so gains from business assets can be rolled-over by the purchase of forestry land (but not standing timber), thereby postponing any CGT liability.

In order to obtain full roll-over relief the value of the new business asset must exceed the sale value of the old asset, however, the cost of the new asset can be increased by subsequent capital expenditure on road, drainage, etc. thereby helping address any initial shortfall in values.

Corporation Tax : Where companies own woodland which is independent of their trading operations, there is no liability to corporation tax in respect of income realized through the sale of growing timber or payments received under forestry grant schemes as outlined above.

Value Added Tax (VAT) : Expenditure on forestry operations will normally attract charges for VAT, so it is generally advantageous to register for VAT purposes. The present threshold for compulsory VAT registration is where taxable supplies exceed £68,000 (2009/10) however, voluntary registration is also permitted. Despite the long-term nature of forestry, which often involves long periods of expenditure without receipts, HM Revenue & Customs will normally accept voluntary registration on the basis that there is a firm intention to produce taxable supplies at a future date.

CENVAT: Central Value Added Tax

Professional tax : It is the tax charged by the State Governments in India. Anyone earning an income from salary or anyone practicing a profession, such as chartered accountant, lawyer, doctor, etc. are required to pay this tax.

Stamp Duty Land Tax : The rates application to commercial property purchases including forestry are as follows:

Price	Rate
£0 - £150,000	0%
£150,001 - £250,000	1%
£250,001 - £500,000	3%
£500,001 and over	4%

Forest Taxation

Every owner of forestland should understand the different taxes that apply to forest ownership and forestry operations, including state and federal income taxes, property tax, and yield taxes (tax applied at the time of harvesting timber).

Income Taxation

Taxation of income derived from forestry investments is a complex topic at both the federal and state levels. Tax laws, regulations, and guidelines are many, detailed and technical. Tax legislation is subject to frequent change and judicial interpretation. The landowner is well advised to seek the advice of a professional tax consultant familiar with timber taxation.

Income taxes on timber revenues are complicated, especially in regards to capital gains treatment. Since income taxation was established in 1913, timber assets have gone from almost no treatment, as capital gains to nearly full treatment.

Capital Gains

Since timber is a long term investment, it is a capital asset subject to capital gains treatment in income taxation. Capital gain is the difference between the selling price of a capital asset and its original cost, or basis. A capital asset is property held by the tax-payer. Capital loss occurs, when the cost of an asset is greater than its selling price. The conditions under which forestry investments and income and loss are considered capital or ordinary.

Timber for capital gains purposes includes standing trees used for wood products. It also includes Christmas trees older than 6 years of age from seed. Timber does not include logs already cut, tops, limbs, stumps, chips, seedlings grown for transplanting, or live trees used for ornamental purposes.

The Timber must be held for a period of 18 months or longer (as measured from the date of acquisition to the date of disposal) to be eligible for long-term capital gains treatment. For landowners who purchase their property as an investment, infrequent lump sum sales of timber are allowed capital gains treatment. In a lump sum sale, payment is made for the standing timber as a whole unit and not on a per unit cut basis. Payment is usually made before the timber is cut. These sales must be discontinuous and isolated. If the tax-payer makes a substantial effort to promote the sale, capital gains treatment might be denied. This type of sale is generally for owners who are not managing their property for timber.

Section 631b of the Internal Revenue Code (IRC) is of special interest to forest landowners who actively manage their forest as a business. This section affords capital gains treatment to the tax-payer who retains an economic interest in the timber. The seller must retain legal title to the timber until it is cut. The date of cutting is when the volume of timber is first accurately determined. The seller must be paid on a per unit cut basis. When selling timber on the stump, your contrast must state that you retain title to the timber until it is scaled and that you are paid on a per unit basis. A well written contract is essential to demonstrate a retained economic interest in the timber.

Depletion Allowance

As timber is cut, the original capital investment, the basis of the timber, is reduced or depleted. Because the trees are rarely all cut at the same time, the original cost must be modified to give the adjusted basis. The adjusted basis is the original purchase price of the timber adjusted for addition or deletion of the capital of the property. The depletion allowance is deducted from timber sale receipts in calculating taxable income. In this way the original investment in timber is recovered.

Record keeping

Up-to-date records must be kept to accurately assess tax liability and take advantage of tax reduction provisions. Two basic accounting methods may be used: Accrual accounting (transactions are recorded) and Cash accounting (when cash is paid).

The timber account is usually divided into sub-accounts for plantation, sub-merchantable timber, and timber. The plantation subaccount documents the costs of site preparation, seedlings, planting, planting tools, and some costs of releasing the plantation. As the seedlings reach the sapling stage, the plantation sub-account is closed and amounts are transferred to the merchantable timber subaccount. Likewise, when the timber becomes merchantable, it is transferred to the merchantable timber subaccount. For mature timber, both cost and volume data should be kept.

Expenses

Expenses may be capital, ordinary, or carrying charges. Capital expenses are related to obtaining or improving a capital asset. They are not deductable in the year they are incurred but are recovered when the capital asset is sold, that is, when timber is harvested. The expenses of a sale are costs that must be charged against capital gains.

Reforestation Tax Credit

The following example illustrates the procedure used to calculate the reforestation tax credit.

Assume a small landowner has 12,000 USD inn eligible reforestation expenses, of which 75% is covered by cost – share programs. Under the current law, 10% of the amount, exclusive of the cost – shared amount, is eligible for the reforestation tax credit.

The reforestation tax credit amount

= 12,000 USD – (0.75 x 12000 USD) (the cost-share grant amount)

= 12,000 USD – 9000 USD

= 30,000 USD

The landowner can take a 10% tax credit on the 3000 USD invested; 3000USD x 0.10 = 300 USD.

This is a tax credit that reduces your tax liability, not a deduction from income.

In addition, the full 3000USD can be amortized over a 7 years period. Because the investment is assumed to occur in mid-year, only $1/14^{th}$ may be deducted in years 1 and 8. In years 2 to 7 the landowner deducts $1/7^{th}$ of the investment. This deduction for the 3000 USD investment is calculated as follows:

Years 1 and 8: 3000 USD ÷ 14 = 214.28 USD

Years 2 through 7: 3000 USD ÷ 7 = 428.57 USD

Property Taxation

The traditional means of revenue generation for local governments in California has been the collection of an annual property tax based on the current assessed value of property at its highest and best use. On forest property, where annual taxes must be paid out of infrequent income from timber sale, annual taxes have been a disincentive to sustainable forest management for small landowners. The ad valorem tax (property/real estate value tax) has often been seen as an incentive for converting the land to its defined "highest and best use", which is usually the conversion of open space forest to subdivision and homes.

In 1976, the State Forest Taxation Reform Act established a timber yield tax and the timberland preserve zone, later renamed timberland production zone (TPZ), to emphasize the timber production and working forest intent of the legislation. The annual property tax on standing timber value was changed to a yield tax on harvested timber.

Timberland Production Zone

Timberland Production Zone (TPZ) is a county zoning classification in which the primary allowed use is the production of timber. Annual property tax is paid, but the assessed value of TPZ land is based on its value for growing timber, not its development value for homes. This usually results in a lower tax assessment than customary "highest and best use" valuation. In return for the lower assessment, the property must be managed for timber and compatible uses for a minimum of 10 more years.

TPZ property valuation is determined in a joint process between the country assessor and the state board of equalization. TPZ land is put into five productivity classes by the county assessor corresponding to timber site class (Sites I to V) as determined from soil and vegetation surveys or other sources. The state board of equalization prepares timberland value tables based on these productivity classes. These tables show the bare land value for TPZ land. The assessor can add to these values for existing compatible uses, such as grazing value per acre.

There are two ways to obtain TPZ zoning if it is not currently applied to a property. Under the first procedure, the landowner petitions the county to classify the land as TPZ. The land must be capable of growing and harvesting timber, and it must meet the following criteria:

1. A forest management plan prepared or approved by a registered profession forester (RPF) must be submitted. Timber harvest must be provided for in a reasonable time.
2. The parcel must meet stocking requirements set forth in the forest practice rules adopted by the State Board of Forestry for the district in which the parcel is located. Alternatively, the landowner can make an agreement with the county that stocking requirements will be met within 5 years.
3. At the county's option, two additional requirements may have to be met. These are a minimum contiguous ownership of up to 160 acres (some counties elect to make this contiguous ownership smaller) and that the land must be Site III or better.

The second way is that timberland may be added to an existing TPZ parcel, if the landowner already has TPZ land contiguous to the new property. Some countries require physical contiguity, while others require the addition to the managed jointly, without physical proximity.

Yield Taxation

Under the current yield tax system, a tax is paid on the value of the timber just before it is harvested. This is ***the stumpage value****, the value of the tree standing on the stump.* The owner of the timber at the time it is cut, is responsible for paying the tax. Often the Licensed Timber Operator (LTO, or logger) who harvests the timber arranges for this tax to be paid.

Tax-payers use the current Harvest Value Schedule to compute the yield tax. Every six months, the State Board of Equalization, Timber tax Division, publishes the Harvest Value Schedules that show average timber prices for geographical areas called timber value areas (TVAs)

Summary

Tax laws are complex, and most forest land-owners will require the assistance of qualified tax professionals, if they have relatively active management programs involving timber production. A few general principles can help you and your family.

- Keep accurate, well organized, detailed records. If you are engaged in timber production, keep sub-accounts for land, timber, buildings and improvements, and Christmas trees. In many cases you will also want to keep separate subaccounts for plantation documents (site preparation, planting, etc.), sub-merchantable timber, including volume data for older stands.

- Reporting your timber sale income as capital gain rather than as ordinary income can reduce taxes by avoiding the 15% self employment tax. Capital gain income also has a lower marginal tax rate than ordinary income. Spread income over tax years, if possible.
- Many costs of managing and maintaining your forest land are tax deductible.
- Tax credits (a direct reduction in income tax) for reforestation expenses are available for 10% of allowable costs. These credits, upto $10,000 per year, are available for seedlings, planting, and other expenses including brush removal upto 2 years after planting.
- Tax deductions (reducing taxable income) for reforestation expenses are available in addition to the tax credits. These deductions are spread over 7 years.
- Generally, you will want to expense costs (deduct them in the year they occur), depreciate what cannot be expensed, and add to the basis what cannot be expensed or depreciated, to reduce the capital gain.
- Losses from disasters such as fires or windstorms are deductible as a casualty loss.
- Cost-share payments must be reported as income, though you may be able to ultimately exclude some or all of those payments.
- Remain an active participant in your business.
- Demonstrate a profit motive in your forest property ownership by having a written long-term management plan. This helps demonstrate your profit motive so that the IRS will not consider your activity a hobby.

Whether you manage for timber production or not, it is extremely important to protect your estate. You may also want to consider the benefits of entering into conservation easements as a way of reducing tax burdens and ensuring that your management style is carried forward.

❒❒❒

Chapter 12

Fundamentals of Project Planning and Management

What is Project Management?

Project management is the application of knowledge, skills, tools, and techniques to project activities to meet project requirements.

Project management is accomplished through the use of the following 5 processes:

1. Initiation
2. Planning
3. Execution
4. Controlling
5. Closure

The project team manages the various activities of the project, and the activities typically involve:

1. Competing demands for: scope, time, cost, risk, and quality.
2. Managing expectations of stakeholders.
3. Identifying requirements.

It is important to note that many of the processes within project management are iterative in nature. This is partly due to the existence of and the necessity for progressive elaboration1 in a project throughout the project lifecycle; i.e., the more you know about your project, the better you are able to manage it. The term **"project management"** is sometimes used to describe an organizational approach to the management of ongoing operations. This approach treats many aspects of ongoing operations as projects to apply project management techniques to them. A detailed discussion of the approach itself is outside the scope of this document.

Fundamentals of Project Planning and Management

1. **Project Integration Management** describes the processes required to ensure that the various elements of the project are properly coordinated. It consists of project plan development, project plan execution, and integrated change control.
2. **Project Scope Management** describes the processes required to ensure that the project includes all the work required, and only the work required, to complete the project successfully. It consists of initiation, scope planning, scope definition, scope verification, and scope change control.
3. **Project Time Management** describes the processes required to ensure timely completion of the project. It consists of activity definition, activity sequencing, activity duration estimating, schedule development, and schedule control.
4. **Project Cost Management** describes the processes required to ensure that the project is completed within the approved budget. It consists of resource planning, cost estimating, cost budgeting, and cost control.
5. **Project Quality Management** describes the processes required to ensure that the project will satisfy the needs for which it was undertaken. It consists of quality planning, quality assurance, and quality control.
6. **Project Human Resource Management** describes the processes required to make the most effective use of the people involved with the project. It consists of organizational planning, staff acquisition, and team development.
7. **Project Communications Management** describes the processes required to ensure timely and appropriate generation, collection, dissemination, storage, and ultimate disposition of project information. It consists of communications planning, information distribution, performance reporting, and administrative closure.
8. **Project Risk Management** describes the processes concerned with identifying, analyzing, and responding to project risk. It consists

of risk management planning, risk identification, qualitative risk analysis, quantitative risk analysis, risk response planning, and risk monitoring and control.

9. **Project Procurement Management** describes the processes required to acquire goods and services from outside the performing organization. It consists of procurement planning, solicitation planning, solicitation, source selection, contract administration, and contract closeout.

❑❑❑

Chapter 13

Government Programmes and Regulations for Agribusiness

Government Programmes for Agribusiness

Both the central government and state governments have devised programmes to woo the entrepreneurs for setting up agribusiness in India. They may be classified as programmes falling under small-scale industries, Khadi village industries, small and medium industries, large industries based on the finance required for investment.. Apart from these to harness the export market after signing the WTO agreement. They include conversion of Export Processing Zones (EPZ) into Special Economic Zones (SEZ), establishment of Agri-export zones. For which government is attracting private investors to make investment in infrastructure development like cold storage chains, improvement in road, rail, sea and air transport systems.

The Ministry of Small Scale Industries & Agro and Rural Industries designs and implements the policies through its field organizations for promotion and growth of small and tiny enterprises, including the coir industries. The Ministry also coordinates with other Ministries/ Departments on behalf the Small Scale Industries (SSI) sector.

The implementation of policies and various programmes/schemes for providing infrastructure and support services to small enterprises is undertaken through its attached office, namely the Small Industry Development Organization (SIDO), statutory bodies/other organizations likely Khadi and Village Industries Commission (KVIC) & Coir Board a Public Sector Undertaking - National Small Industries Corporation (NSIC) and three training institutes - National Institute of Small Industry Extension Training (NISIET), Hyderabad, National Institute for Entrepreneurship and Small Business Development (NIESBUD), New Delhi and Indian Institute of Entrepreneurship (IIE), Guwahati.

Entrepreneurial Training institutes

There are three Institutes engaged in training of small-scale entrepreneurs. These are Indian Institute of Entrepreneurship (IIE), Guwahati, National Institute of Small Industry Extension Training (NISIET), Hyderabad, National Institute for Entrepreneurship and Small Business Development (NIESBUD), New Delhi.

Fragrance & Flavour Development Centre (FFDC), Kannauj

Fragrance & Flavour Development Centre (FFDC) has been set up as an autonomous body in the year 1991 by Govt. of India with the assistance of UNDP/UNIDO and Govt. of U.P/UNDP/UNIDO has provided technical expertise and imported equipments. Govt. of U.P has provided land and building while Govt. of India has been contributing for indigenous equipments and recurring expenditure. Main objectives of the Centre is to serve, sustain and upgrade the status of farmers and industry engaged in the aromatic cultivation and its processing, so as to make them competitive both in the local and global market.

Export Processing Zones

The export zones (EPZ) set up as enclaves, separated from the Domestic Tariffs Areas by fiscal barriers, are intended to provide a competitive duty free environment for export production. There are four EPZs set up by the Government at Noida(Uttar Pradesh), Chennai (Tamil Nadu), Palta (West Bengal) and Vishakhapatnam (Andhra Pradesh).

Special Economic Zones

A new scheme for setting up of Special Economic Zones (SEZs) in the country to promote exports was announced by the Government in the Export and Import Policy on 31st March, 2000. The policy provided for setting up of SEZs in the public, private, joint sectors or by State Governments. It was also announced that some of the existing Export Processing Zones would be converted into SEZs. Accordingly, the Government has issued notification on 1 November, 2000 for conversion of the existing Export Processing Zones at Kandla (Gujarat), Santa Cruz(Maharashtra) and Cochin (Kerala) into SEZs. Notification has also been issued for conversion of the private sector EPZ at Surat

(Gujarat) into the Special Economic Zone at the request of the promoters.

Export Oriented Units (EOU)

The export Oriented units(EOU) scheme introduced in the early 1981, is complementary to the EPZ scheme. It adopts the same production regime but offers a wide option in locations with reference to factors like source of materials, ports of export, hinter land facilities, and availability of technological skills, existence of industrial base and the need for a large area of land for the project. And 1,536 units are in operation under the EPU scheme as on March, 2001.

Product range : EOUs are mainly concentrated in textiles and yarn, food processing, electronics, chemicals, plastics, granites and minerals/ores. Majority of units are located in Tamil Nadu, Andhra Pradesh, Karnataka, Maharashtra and Gujarat.

Export Promotion Industrial Park (EPIP) Scheme : A centrally sponsored " Export Promotion Industrial Park (EPIP) Scheme has been introduced in 1993-94 with a view to involving the State Governments in the creation of infrastructure facilities for export oriented production. The scheme provides that 75 % of the capital expenditure incurred towards creation of such facilities, ordinarily limited to Rs. 10 crores in each case, will be met from a central grant to the State Governments.

New Anna Marumalarchi Thittam: The State Government of Tamil Nadu has introduced the New Anna Marumalarchi Thittam in April 2002 to set up agribusiness units with a minimum investment of Rs 35 lakhs at the rate of one unit in each block. This scheme provides scope for setting 385 agribusiness units in Tamil Nadu.

Agri Clinic and Agribusiness Centres: Small Farmers Agribusiness Consortium in co-operation with MANAGE has drawn plans to provide training on management capacity building for those willing to set up Agriclinics and Agribusiness Centres either as individual or a group five (four agricultural and allied graduates and one management graduate) with a maximum loan assistance of Rs. 10 lakhs for individuals and Rs 50 lakhs for a group of five entrepreneurs.

Regulations

Legal regulations

Decisions are strongly affected by laws pertaining to competition, price setting, distribution arrangements, advertising, etc. It is necessary for a manager to understand the legal environment of the country and the jurisdiction of its courts. The following laws affecting business in India are important :

1. Indian Contract Act, 1872
2. Factories Act, 1948
3. Minimum Wages Act, 1948
4. Securities contracts Regulation Act, 1956 (Now replaced by SEBI Act)
5. The Companies Act, 1956
6. Trade and merchandise Marks Act, 1958
7. Monopolies and Restrictive Trade Practices Act, 1969
8. The water (Prevention and Control of Pollution) Act, 1974
9. The Air (Prevention and Control of Pollution) Act, 1981
10. Sick Industrial Companies (Special provision) Act, 1985
11. Environment protection Act, 1986
12. Consumer protection Act, 1986
13. Securities and Exchange Board of India Act, 1992
14. Taxation laws covering Corporate tax, indirect taxes like Excise, Customs, Sales tax and Wealth tax)

Food Safety and Quality

Fruit Products Order (FPO), 1995

Fruit Products Order-1955, promulgated under Section 3 of the Essential Commodities Act, 1955, aims at regulating sanitary and hygienic conditions in manufacture of fruit, vegetable products. It is mandatory for all manufacturers of fruit, vegetable products to obtain a license under this Order. To ensure good quality products, manufactured under hygienic conditions, the Fruit Product Order lays down the minimum requirements for:

1. Sanitary and hygienic conditions of premises, surrounding and personnel.
2. Water to be used for processing.
3. Machinery and equipment.
4. Product standards.

Besides this, maximum limits of preservatives, additives and contaminants have also been specified for various products.

This order is implemented by Ministry of Food Processing Industries through the Directorate of Fruit & Vegetable Preservation at New Delhi. The Directorate has four regional offices located at Delhi. The Directorate has four regional offices located at Delhi, Mumbai, Kolkata and Chennai, as well as sub-offices at Lucknow and Guwahati. The officials of the Directorate undertake frequent inspections of the manufacturing units and draw random samples of products from the manufactures and markets which are analyzed in the laboratories to test their conformity with the specifications laid under FPO.

The Central Fruit Advisory Committee comprising of the officials of concerned Government Departments, Technical experts, representatives of Central food Technology Research Institute, Bureau of Indian standards, Fruits and Vegetables Products and processing Industry, is responsible for recommending amendments in the Fruit Product Order, In view of the demands of the industry, and the liberalized economic scenario, major amendments were made in FPO during 1997.

Meat Food Products Order (MFPO)

Meat Food Products Order, 1973 (MFPO) promulgated under the provisions of Essential Commodities Act, 1955 provides for sanitary and other requirements, limits of heavy metals, preservatives, insecticides, residue, etc., for meat food products. This order was being implemented by Ministry of Rural Development in the Ministry of Rural Area & Employment. As on 31st March, 1998, there were 128 licenses issued under MFPO 1973.As per the recent amendment to the Allocation of Business, Ministry of Agriculture (Deptt. of Agriculture & Co-operation) would now be the Administrative Ministry for this Order.

Milk & Milk Products Order (MMPO)

Milk and Milk Products Order, 1992 administered by the Department of Animal Husbandry & Dairying under Ministry of Agriculture was promulgated on 9th June, 1992 under the provision of Section 3 of the Essential Commodities Act, 1955 with a view to maintain an increased supply of liquid milk of desired quality to the general public. This order regulated production, supply and distribution of milk and milk products throughout the country. The order also seeks to ensure the observance of sanitary requirements for dairies, machinery and premises, and quality control standards for milk and milk products. So far, 254 registration certificates under MOP, 1992 has been issued by the Department of Animal Husbandry and Dairying.

Codex Alimentarius

The term Codex Alimentarius is taken from Latin and means food code. Codex Alimentarius brings together all the interested parties - scientists, technical experts, governments, consumers and industry representatives to help develop standards for food manufacturing and trade. These standards, guidelines and recommendations are recognized worldwide for their vital role in protecting the consumer and facilitating international trade. As Codex Alimentarius represent a consensus of food and trade experts from around the world, these standards are more and more being used in international trade negotiations and also for settling of disputes by WTO

The Codex contract Point in India is the Directorate General of Health Services (DGHS) in the Ministry of Health; however, the Ministry of Food processing Industries is closely associated with the activities of Codex Alimentarious. Under the Plan schemes, a scheme for setting up of Codex Monitoring Cell in the Ministry with the allocations of Rs. 58.00 lakhs in 1998-99 has been formulated for creating data base, technical examination of various standards in association with experts. A delegation led by a senior officer of this Ministry Participated in the meeting of the Codex Committee on food Labelling at Ottawa from 25th to 29th May, 1998. Another Officer from this Ministry was Member of the Indian delegation for the meeting of the Codex Committee on General Principles held at Paris, France from 7th to 11th September, 1998.

The Pulses, Edible Oilseeds and Edible Oils (Storage) Order. 1977 empowers the government to put maximum stock limits on wholesalers and retailers of pulses, oilseeds and oils and is designed to maintain supplies and ensure equitable distribution and availability at fair prices of these items.

Food Quality

The Food Processing Ministry cleared a proposal for release of Rs.59.2 lakhs to Food Research and Analysis Center, New Delhi for upgradation of its Food Analysis and Quality Control Laboratory for analysis of food products. The main objective of the proposal is to upgrade the existing analytical laboratory and bring it on par with any other modern analytical laboratory in the country. During the year under review, the Ministry cleared a proposal in principle for release of 12.32 lakhs to CCS Haryana Agriculture University, Hissar for upgradation of quality control and food analysis laboratory of the Department of Food Science and Technology of the university. The main objective of the proposal are to provide quality assurance and analytical services to the food processing industries, to undertake micro biological examination of various pathogen and mycotoxins and to estimate nutritional parameters including minerals, vitamins, food value in calories, protein carbohydrates, fats, etc.

Hazard Analysis and Critical Control Point (HACCP)

Hazard Analysis and Critical Control Point (HACCP) is an important quality assurance system. This system ensures that the products are safe and of good quality. The system is extremely desirable in view of the changing scenario in the International trade. The Ministry provides grant of 50% subject to a limit of Rs.10 lakhs towards the cost of implementing Total Quality Management (TQM) including HACCP and ISO-9000 certifications.

This Ministry sponsored a one-day seminar and five days training programme organized by APEDA from 30th November to 5th December, 1998 in collaboration with NSF-International strategic Registration Limited, USA, which is the main authority for certifying HACCP-9000. HACCP is an important requirement for ensuring the quality of products from health and safety aspects and is crucial for exports.

Laws relating to food processing industries

There are a number of food laws being implemented by various Ministries/Departments. These are primarily meant for two purposes, namely :

1. Regulation of Specifications of food and
2. Regulation of Hygienic condition of Processing/Manufacturing.

Some of these food laws are mandatory and some are voluntary. The details of various food laws in operation in India are as under:

A. Food Laws

1. Prevention of Food Adulteration Act (Ministry of Health)

The Act lays down specifications for various food products and is mandatory. The Ministry of Health in 1995 had constituted a Task Force under the chairmanship of Shri E.S. Venkataramaiah, Chief Justice of India (retired). The Task Force recommended that there should be emphasis on good manufacturing practices instead of detection of adulteration and prosecution. It also expresses concern about lack of laboratory equipments and quantified persons. In addition, it is also suggested that the name of PFA Act be changed to Food Safety Act.

2. Agriculture Produce (Grading & Marking) Act (Ministry of Rural Development)

This Act is commonly known as AGMARK and is voluntary. The Act lays down the specifications for various agricultural commodities including some processed foods.

3. Laws being operated by Bureau of Indian Standards (BIS)

BIS is the largest body for formulating standards for various food items. These standards are also voluntary.

4. Essential Commodities Act

A number of quality control orders have been issued under Essential Commodities Act, such as FPO, MMPO, Meat Product Order and Vegetable Oils Control Order. These orders are mandatory and primarily meant for regulating the hygienic conditions. They need to be clubbed under one order which may be called Food Products Order.

B. Harmonization of Food Laws

The review of multiple laws is necessary to have a uniform and logical approach for regulating the quality of food. The following action is being taken by various Ministries:

1. The Ministry of Civil Supplies & Consumer Affairs has brought out a paper for consideration of Committee of Secretaries (COS). The paper recommends that BIS should formulate standards for all food items in the country. This will be a major step towards harmonization of food laws and is still under consideration of COS for finalization.

2. The Task Force constituted by the Prime Minister under the chairmanship of Shri Nulsi Wadia has submitted its report which is under the consideration of the Government. The Task Force had advocated promotion of food safety and quality. The Task Force has further made following suggestions :

   - Food Regulation Authority (FRA) is set upto formulate and update food standards for domestic and export market.

   - FRA should replace the PFA to conform to international standards. The Task Force has given ten specific recommendations, such as provision of storage simplicitor, simplification of sampling procedure, simplification of procedure for nominee, time limit for prosecution, standard methods of analysis to be prescribed, penalty should graded according to the gravity of offences & provision of adequate/infrastructure and laboratories.

   - Harmonization of Indian standard with quality norms of Codex and WTO.

   - The Central Committee of food Standard (CCFS) should be replaced by FRA Governing Body for expeditious decisions.

Whom to approach for what?

S. No.	Steps involved	Persons/ Officers to be contacted
1	Product selection and assistance in preparation of project report/feasibility/report/market survey	Small Industries Service Centre, Chennai
2	Obtaining provisional /permanent registration certificate	District Industries Centre of the concerned district where the unit is to be set up
3	Registration under sales tax	o/o The Commissioner of Commercial Taxes, Chennai
4	Registration under central excise	O/O The Collector of Central Excise & Customs, Chennai
5	Obtaining Finance	Nationalized banks Private Commercial banks Tamil Nadu Industrial Investment Corporation Small Industries Development Bank of India
6	Obtaining water supply	Tamil Nadu Water Supply & Drainage board
7	Obtaining Power connection	Tamil Nadu Electricity Board
8	Registration under ESI Act (applicable for units employing 20 or more workers)	Employees State Insurance Corporation, Chennai
9	Obtaining ISI Certificate	Bureau of Indian Standards, Nanak Bhavan, New Delhi
10	Obtaining the Trade mark registration	Registrar of Trade markers, 9 Judge jumbulingam Mudaliar road, Chennai
11	Obtaining patent Right	The Controller of patents and Design, 776 Triplicane high road, Chennai
12	Obtaining license under food products order for manufacture of food products	Deputy director, Directorate of Marketing 7 Inspection, Ministry of Food & Agriculture, Shastry Bhavan, 35 Haddows Road, Chennai-600006.
13	Obtaining Information on manufacturing of essential oils	Central Institute of Medicinal 7 aromatic plants, Regional Centre, UAS, GKVK Campus, Hebbal, Bengaluru
14	License for Drug &Cosmetic manufacture	Drug Controller Office of the Drug Controller, Chennai
15	Pollution control License	Tamil Nadu Pollution control Board, Chennai
16	Setting up of Rice Solvent Extraction Units	Tamil Nadu Civil Supplies Department, Chennai

Agricultural Inputs and Agro-processing

Agro-inputs are either biological or chemical or inorganic compounds used in the production of agricultural and allied products. Some of the agro-inputs used in India is :

1. Inorganic fertilizers
2. Pesticides which includes insecticide, fungicide, nematicide and herbicide
3. Seeds which includes varieties, hybrids and genetically modified plant materials
4. Vegetatively propagated planting materials
5. Bio-fertilizers
6. Organic inputs which includes compost, vermi-compost, enriched farm yard manure, oil meals, farm yard manure
7. Bio-pesticides
8. Bio-control agents
9. Plant Growth regulators
10. Micro-nutrients
11. Farm machinery and agricultural tools and implements
12. Animal feeds
13. Poultry feed

The production and distribution of above mentioned agricultural inputs creates agribusiness opportunities. The agricultural input production inputs and distribution firms may be big companies like Madras Fertilizers, SPIC, FACT, EID Parry Monsanto, Syngenta or MAHYCO and the like. They may also be small business firms like vermi-composting, coir pith decomposing.

The agro and food processing industry in India is utmost significance in terms of employment and income generation, poverty alleviation, export promotion and foreign exchange earnings. The growth of the processed food sector is expected to make a quantum jump of Rs.1,75,000 crores by 2005. Although India is one of the largest producers of raw materials for the food processing industries in the world, the industry itself is underdeveloped in India. Less than 2 per cent of fruits and vegetables production is processed, compared to 30 per cent in Thailand, 70 per cent in Brazil, 78 per cent in Philippines and 80 per cent in Malaysia. The value addition in food sector is as low as 7 per cent. There is need for increasing food processing from 2 to 10 per cent by 2010. This will require an investment of Rs. 1,40,000 crores in the food processing sector. The investment will generate direct employment for 77 lakh persons and indirect employment for three crore people. This will reduce wastage by Rs.80,000 crores. Apart from these advantages, the value addition of food product will go up from 7 to 35 per cent which will be reflected in the corresponding increase in Gross National Product (GNP).

Types of food processing industries

1. Primary processing industries
2. Rice milling
3. Roller Flour milling
4. Pulse milling
5. Oil milling
6. Cereal based products (bread-biscuits)
7. Cocoa Products
8. Soft drinks
9. Alcoholic drinks
10. Horticulture
11. Fruits and vegetables
12. Milk and milk products
13. Meat processing
14. Poultry processing
15. Fish processing
16. Food packaging

A strategic Assessment of Food, Beverage and Agribusiness Opportunities in India are furnished in Table 1.

Table 1. Processed Food markets

Product	Volume (metric tonnes)	Value (US $ million)	Major players
Basic foods			
Packaged wheat flour (including branded flour)	1 million	8	Hindustan Lever Limited(HLL), Pillsbury, Best foods India Ltd.,
Spices	2.45 million	3780	Regional
Edible oil	8 million	8820	ITC agrotech, Marico Industries, HLL, NDDB, Ahmed Oil Mills Ltd
Salt	500,000	60	Gujarat Salt Federation, Hindustan Salt Works, Tata Chemicals, Bestfoods India Ltd.,
Sugar	15 million	7,150	Government, co-operatives
Eggs	33 billion eggs	780	Government Poultry Corporations, National Egg Co-ordination Committee
Poultry meat	550,000	n.a.	Govt. Poultry Corporations
Milk (total)	73.5 million		NDDB
Bakery products(including unorganized sector)			
Biscuits and cakes	1 million (biscuits) 500,000(cakes)	775	Britannia Industries Ltd.,, Parle Products, Bakeman's
Bread	1.5 million	350	Britannia India Ltd.,Modern Industries (now with HLL), Spencer's
Indian Dairy products			
Ghee (Butter oil) (organized sector only)	85,000	n.a.	
Indian Milk sweets(including unorganized sector)	300,000	350	Small players
Western dairy foods (organized sector)			
Ice cream	50 million liters	120	HLL, Gujarat Co-operative Milk Marketing Federation, Hutsun Agro
Butter	80,000	200	Gujarat Co-operative Milk Marketing Federation, Britannia Industries Ltd.,
Cheese	12,000	30	Gujarat Co-operative Milk Marketing Federation, Britannia Industries Ltd.,

Product	Volume (metric tonnes)	Value (US $ million)	Major players
Dairy whitener	48,000	85	Nestle India Ltd., Gujarat Co-operative Milk Marketing Federation
Western dairy foods (organized sector) contd..	100,000	100	Gujarat Co-operative Milk Marketing Federation, Nestle India Ltd.,
Malted food drinks	65,000	140	Cadbury India Ltd., Smithkline Beecham Consumer products, Nestle India Ltd.,
Processed Fruit and Vegetable Products			
Pickles	120,000	28	Nestle India Ltd., American Dry Fruits, Ruchi, Bedeker's, Priya
Fruit Beverages	130,000	80	HLL, Parle Products, Enkaye,Texofoods
Fruit spreads, sauces & ketchup	50,000	43	HLL,Marico Industries
Convenience foods (organized sector)			
Chocolate	20,000	95	Cadbury India ltd., Nestle India Ltd.,
Sugar-boiled confectionery	80,000	280	Parry's Confectionery, Nestle India Ltd.
Chewing /Bubble gum	25,000	55	Warner Lambert, Wrigley's Perfetti
Snack foods			
Traditional Indian snacks (incl. small players)	310,000	295	Haldiram's. Pepsi Foods
Western snacks (including branded potato chips = 5,000 metric tonnes)	40,000	30	Frito-Lay India, Uncle Chipps, Procter & Gamble

Source: Promar International, 2001. Indian Food Industry, 20(4): 29.

❑❑❑

Chapter 14

Practical Exercises

1. Economic Analysis of Arecanut Based Agroforestry Systems
2. Model Project Preparation on Raising of Eucalyptus Plantation
3. Computation of Depreciation by Different Methods
4. Net Worth Statement of Forestry Sector

Exercise-1

1. Economic Analysis of Arecanut Based Agroforestry Systems

1. Spacing and alignment of arecanut and associated crops in different mixed cropping systems

Components of mixed cropping systems	Spacing	Plant population/ha	Total plants/ha	CII
A alone	2.7m x 2.7m	1372	1372	1.00
A+B	2.7m x 2.7m 5.4m x 5.4m	1372 343	1715	1.15
A+P	2.7m x 2.7m 2.7m x 2.7m	1372 1372	2744	2.00
A+B+P	2.7m x 2.7m 5.4m x 5.4m 2.7m x 2.7m	1372 343 1372	3087	2.15
A+C	2.7m x 2.7m 2.0m x 1.5m	1372 3333	4705	2.08
A+C+P	2.7m x 2.7m 2.0m x 1.5m 2.7m x 2.7m	1372 3333 1372	6077	3.08
A+C+B	2.7m x 2.7m 2.0m x 1.5m 5.4m x 5.4m	1372 3333 343	5048	2.23
A+C+B+P	2.7m x 2.7m 2.0m x 1.5m 5.4m x 5.4m 2.7m x 2.7m	1372 3333 343 [illegible]	6420	3.23

A – Arecanut, B – Banana, C – Cardamom, P – Pepper

The population of sole banana crop – 2279

The population of sole cardamom crop – 3086

2. Establishment expenditure of arecanut plantation (as per package of practices)

Particulars	Cost (Rs.)/ha
1.Land rent/lease value for six initial years of establishment @ **Rs.5000**/year for 6 years	30,000
2. Land preparation for planting	
a. Removal of stones, stubs, etc. and levelling	10,000
b. Preparation of drainage channels including main drains for total plants @ Rs ___ / plant (minimum of 3' to 3.5' deep)	13,700
c. Preparation of pits and planting. Pit size 3'x3'x3' spaced at **2.7x2.7m** @ Rs.**10** per pit and **1370** pits per ha	13,700
d. Fencing around the plantation including digging of cattle proof trench @ Rs.**60** per mtr length for **400** metres.	24,000 10,000
e. Repair and maintenance of fence for first six years.	
3.Planting material	
a. Arecanut seedlings: **1370** seedlings @Rs.**10** /seedling	13,700
4.After care	
a.Nurse crop for initial years for procuring and planting of **1370** bananas @ Rs.**2** /plant.	2740
b. Irrigation cost, Application cost @ **3** labours / Irrigation / ha for initial 6 years of establishment.	28,800
c. Purchase of sprayers, Agricultural equipments including irrigation pipes, electric motors, etc.	25,000
5. Inputs	
I. Fertilizers and manures	
a. Fertilizers: Recommended dose of 100g N, 40g P_2O_5 and 140 K_2O per plant per year from third year onwards and 1/3rd and 2/3rd during I and II year of planting.	15,000 51,375
b. Manures: Compost and green leaf mulch @ 25kg per plant per year for initial 6 years 25kg x **1370** plants x 6 years = **205.50** tonnes cost @ Rs. **250** per ton.	
II. Plant protection chemicals per ha, for initial 6 years, **1370** plants @ Rs.**2** per palm per year.	16,440

6. Labour Cost @ Rs. 250 Operational works	No. of labours	Cost (Rs.)/ha
a. Application cost of PPC	25	6250
b.Application cost of fertilizer and manures @ 50 labours per ha per year for initial 6 years	300	75,000
c.Weeding, l oosening of soil by light digging, repairs of drainage channels etc. @ 75 labours per ha per year for initial 6 years.	450	1,12,500
d.Watch and ward and collection of fallen fronds, replanting and such other works @ 100 labours per year.	600	1,50,000
e. Interest on fixed capital	-	25,000
Total establishment cost		**6,23,205**

Total establishment cost (Rs/ha)

$$\text{Imputed cost} = \frac{\text{Total establishment cost (Rs/ha)}}{\text{Life of arecanut plantation (50 years)}}$$

$$\text{Imputed cost} = \frac{6{,}23{,}205}{50} = \text{Rs.}12{,}464/\text{ha/year}$$

3. Recurring cost on cultivation of arecanut plantation/ha/year

Particulars		Expenditure
1.Imputed value of establishment expenditure		12,464
2.Annual land rent		5000
3.Depreciation on machinery equipments, buildings, etc.		2050
4.Inputs		
a.Plant protection chemicals		6975
b.Fertilizers and manures		
i. Fertilizer @ 100:40:140 gm per palm		4000
ii. Manures, compost and green leaf @ 25 kg per palm		8568
5.Labour cost (both skilled and ordinary until marketing) (Ordinary labours @ Rs. 250 and Skilled labour @ Rs. 300)	**No. of labours**	**Cost (Rs/ha)**
a.Weeding, loosening of soil by light digging, repair of drainage channels, etc.	50	12,500
b.Mulching with green and dry leaves grasses from soppina betta	50	12,500
c. Application of fertilizers and manures	50	12,500
d. Application of Bordeaux mixture (1%) and other plant protection chemicals	40+10	13,000
e. Collection of fallen nuts and watch and ward during cropping season	125	31,250
f. Harvesting and lowering fruit bunches	30+15	12000
g. Dehusking and processing	100	25,000
h. Grading and bagging	10	2500
i. Transport and marketing charges	20	5000
Total yearly cost (Rs/ha)		**1,65,307**

4. Cost of cultivation, processing and marketing of mixed crops in arecanut plantation

Particulars	Cost (Rs/ha) of mixed crops		
A. Establishment cost	**Pepper**	**Banana**	**Cardamom**
1.Land preparation for planting	1890	1715	2000
2.Planting material	2055	686	3334
3.Planting cost	2055	1372	3334
Total cost	**6000**	**3773**	**2668**
B. Recurring costs			
1.Imputed cost of establishment	300	3773	1444
2.After care (weeding, mulching, earthing up, etc.)	5500	3773	1450
3.Inputs per ha			
i. FYM and green leaf mulching	6000	5145	2800
ii. Fertilizers	6250	2500	4500
iii. Plant protection chemicals	7500	5000	3500
4.Harvesting and marketing			
i. Harvesting charges	10015	1000	2300
ii.Cleaning, processing and preparation for marketing	1250	1200	1690
iii. Transportation and marketing charges.	500	2157	2834
Total cost	**37,315**	**24,548**	**20,518**

5. Comparative economics of arecanut based agroforestry systems

Agroforestry system	Intercrops in arecanut	Maintenance cost (Rs/ha)			Arecanut		Mixed crop		Gross Returns	Net Returns	Percent increase over sole crop	B:C ratio
		Arecanut	Mixed crop	Total	Yield (q/ha)	Returns (Rs/ha)	Yield (q/ha)	Returns (Rs/ha)				
A alone	-	165307	-	165307	40.42	808400	-	-	8084[illegible]0	643093	-	1:4.89
A+B	Banana	165307	24548	189855	36.17	723400	69.11 (B)	276440	999840	809985	25.95	1:5.26
A+P	Pepper	165307	37315	202622	39.59	791800	25.47 (P)	101880	893680	691058	7.45	1:4.41
A+B+P	Banana Pepper	165307	24548 37315	227170	37.68	753600	58.86(B) 16.75 (P)	989190	1742790	1515620	135.67	1:7.67
A+C	Cardamom	165307	20518	185825	38.50	770000	1.65 (C)	66000	836[illegible]0	650175	1.10	1:4.40
A+C+P	Cardamom Pepper	165307	20518 37315	223140	42.74	854800	1.28 (C) 22.22 (P)	1051100	1905900	1682760	161.66	1:8.54
A+C+B	Cardamom Banana	165307	20518 24548	210373	38.22	764400	1.05 (C) 55.96 (B)	265840	1030240	819867	27.48	1:4.80
A+C+B+P	Cardamom Banana Pepper	165307	20518 24548 37315	247688	41.79	835800	0.61 (C) 56.76 (B) 15.67 (P)	956590	1792390	1544702	140.19	1:7.20

Market price: Arecnut – Rs. 20,000 / q; Banana – Rs. 4,000 / q; Pepper- - Rs. 45,000 /q; Cardamom – 40,000 / q

Exercise-2

1. Exercise on Model Project Preparation on Raising Eucalyptus Plantation

Eucalyptus is water- intensive and destroys the soils and the underground water table

Studies have shown that Eucalyptus consumed 0.48 litres of water to produce a gram of wood, compared to 0.55, 0.77, 0.50 and 0.88 litre per gram for siris, shisham, jamun and kangi respectively. Thus, Eucalyptus is more water efficient than many indigenous species. However, the mean annual growth of Eucalyptus is about 8 cum/ha - 40 cum/ha, as compared to the average of 0.50 cu m/ha for indigenous trees. Being much faster in growth; the very reason for which the species was introduced in the country; the water and nutrient absorption of the tree is much more than the slow growing indigenous trees.

The drought hardiness of the species comes from the fact that Eucalyptus has deep-rooted system and an ability to absorb water even at higher moisture tension level, than many other mesophytes plants. *E. tereticornis* planted in arid and semi-arid areas of Portugal 15 years back, have shown no evidence of soil degradation over the years.

However, though Eucalyptus is an excellent industrial species, providing timber for poles, pulp and fuelwood, it cannot be used as fodder plant and provide other non-timber uses, limiting its role as a social forestry tree. Thus, plantations of eucalyptus may be limited as industrial plantations with management regime drawn parallel to any intensively produced crop.

Introduction

Eucalyptus belongs to the family Myrtaceae with about 300 species of the genus. The species is one of the fastest growing trees in the world and many species attain great heights. *Eucalyptus amygdalin* is the tallest known tree with specimens attaining a height of as much as 480 feet.

Basically, a native of Australia and Tasmania, Eucalyptus was introduced in India, by the British in 1843, in Nilgiri Hills as an experiment to find high yielding species for fuel and timber. It soon became a favoured species for the foresters/commercial plantations, owing to its fast growth, non-exacting, non-browseable and drought resistant nature and adaptability to a variety of agroclimatic conditions.

Eucalyptus is popularly known as gum tree, red iron tree, nilgiri or safeda. Many fast growing species suitable for commercial cultivation in India have been identified. *Eucalyptus tereticornis* and *E. grandis* are important commercial species with a clean straight bole and compact crown. Hybrid eucalyptus (combination to *E. tereticornis* and *E. grandis /E. urphaylla/ E. camaldulensis*) has shown greater vigour and drought and insect resistant characters in the field conditions and are preferred planting stock for commercial plantations.

Botanical Features

Eucalyptus is a fast growing, medium- sized to tall tree attaining 20-50m in height and upto 2m in diameter. The tree has a deep tap root system with mycorrhizal associations which increases its ability to draw nutrients and water. The tree has a smooth silvery white stem. The leaves are leathery in texture, hang obliquely or vertically and are studded with glands containing aromatic oil. Flowering takes place during July-August. Flowers in bud are covered with a cup- like membrane (whence the name of the genus, derived from the **Greek 'eucalyptos' meaning- 'well covered'**), which is thrown off as a lid when the flower expands. The fruiting occurs during September - October. The fruits are surrounded by a woody, cup-shaped receptacles and contain numerous minute seeds.

Silvicultural Characteristics

Eucalyptus is versatile, fast growing and strongly coppicing tree possessing a wide range of soil and climatic adaptability. *E.tereticornis* has the most extensive latitude range (9-38^{0}S) of any species in the genus. Basically a light demander, the growth of the species is very much reduced under shade. Eucalyptus is known for its drought hardiness, although annual rainfall of 800 mm is preferred. The species is also moderately salt tolerant and relatively fire resistant. Eucalyptus is generally regarded as frost sensitive, though in Uruguay, *E. tereticornis* has known to come up with reasonable success in regions where unseasonal frost is likely to occur. The species is known to suffer cholrosis and die-back due to the reduced iron absorption in alkaline soils.

The species grows under a wide range of climatic/soil conditions from warm to hot, sub humid to humid and from good to degraded soils. The range of agro-climatic conditions of the species (*E. tereticornis*) is given as under:

Climate

- Altitude range : 0-1000 m above sea-level.
- Mean annual rainfall : 500-3000 mm
- Rainfall regime : summer, winter, bimodial: uniform
- Dry season duration : 0-8 months
- Mean annual temperature : 10-27°C
- Mean maximum temperature of hottest month : 22-42°C
- Mean maximum temperature of coldest month : - 2-19°C
- Absolute minimum temperature: > -8°C (data for upper limit NA)

Soil and Physiography

- Soil texture : light, medium, heavy
- Soil drainage : Free, seasonally water-logged
- Soil reaction : acidic, neutral
- Special soil tolerances : saline
- Soil types : alluvial soils, gravely soils, ferral soils, red soils, sandy soils

Nursery practices

Eucalyptus can be easily propagated from seeds, as well as, through vegetative propagation by cuttings (clonal propagation). On an average, there are about 642,000 viable seeds per kilogram of seed and chaff mix. Dry seeds at 5-8% moisture content, can be stored in air tight containers under refrigerated conditions (3-5°C) for more than 10 years without losing its viability.

Under nursery conditions, seeds are sown on raised beds under shade. Addition of mycorrhiza innoculum to the nursery soils by adding soil from natural eucalyptus forest is highly beneficial for establishment and growth of the plants. No pre-sowing treatment is required. Rapid and complete germination is achieved under moist, warm (25-35°C) conditions in presence of light. Seedlings are pricked out and transferred to polybags at the second leaf-pair stage i.e. about 6 weeks from sowing. Seedlings are planted out in the field, when they reach a height of about 25 cm i.e. about 3-5 months after sowing. This should coincide with the onset of monsoon season.

Plants at nursery stage are highly susceptible to damping off and other fungal diseases, which can be limited by strict attention to hygiene, reducing watering and shade and allowing good ventilation.

Cultivation practices

Eucalyptus, generally raised for industrial plantations - mainly pulpwood, firewood or poles, are maintained with a shorter rotation of 5-7 years. For commercial plantations, intensive site preparation by ploughing or deep ripping on compact sites, is beneficial. On wet sites, moulding should be adopted to improve root aeration and provide well-drained condition that facilitate planting. Spacing adopted is 2m x 2m or 1.5 m x 1.5 m (high density plantations). In case, crops are cultivated between the rows (agroforestry), wider spacing of 4m x 2m ; 6m x 1.5m or 8m x 1m are recommended. Nursery raised seedlings/plantlets in polybags may be planted at the onset of monsoons, in pits of 45 cm x 45 cm x45 cm. Organic manure mixtures along with fertilizers containing 25g of NPK (3:2:1) and 50 g of phosphate should be applied in the planting pit at the time of planting. Protective irrigation is essential, in case of monsoon failure, in the first two years of plantation. Eucalyptus is intolerant to shade and does not compete well with grasses for water and nutrients, thus 2-3 hand weeding and soil working in the initial stages are essential.

Owing to its fast growth, Eucalyptus is a heavy feeder and requires supplements in the form of organic and chemical fertilizers in successive years. Deficiency of Nitrogen in soils is a limiting factor for growth and can reduce the yield by 60%. For maintaining the soil fertility, it is advisable to raise Eucalyptus trees with legumes as an intercrop.

Harvesting is done by clear felling the stand in 6-7 years. Once the tree is felled, the stump throws many coppice shoots. These should be singled out to keep only one vigorous stem per stump, which will form the second crop. It is advisable to change the planting stock after the second harvest, as there is loss in vigour in coppice from the third coppice onwards.

Pest and diseases

One of the most serious diseases of *E. tereticornis* is canker caused by fungus, *Corticium salmonicolour* known as pink disease. Other fungal pathogens known to cause damage include *Ganaderma lucidium, Endothia gyrosa* and *Cylindrocladium spp. Cylindrocladium clavatum* has been recorded to cause seed rot, seedling blight and seedling wilt of *E. tereticornis* in Punjab. Other potentially serious diseases are web blight (*Rhizoctonia solani*) in the nursery and stem canker caused by *Cryphonectria* and *Cytospora eucalyptiocola* which can cause heavy mortality.

Among insects, ceranbycis beetle, *Celstems scabrator* is reported to attack young plants in plantations. Subterranean termites are reported to damage seedlings and young plants of the species.

Yield

The mean annual increment (MAI) in volume over bark of *E. tereticornis* on medium quality sites at age 8-10 years is about 15 cu m/ha. A review of the performance of the species (*E. tereticornis*) from 8 years old plantations, through out India showed a MAI of 1.3 - 19.8 cu m/ha depending on stocking and site quality. Highest wood yield of 105 tonnes/ha in five years has been reported in red sandy clay loam soils, under irrigated conditions in Karnataka.

On an experimental scale, the best provenances on the best sites in Bangladesh yielded over 60 cu m/ha/year after five years, at a planting rate of 10,000 stems/ha. Under special trials in China, MAI at 5.5 years of age for *E. tereticornis* was 11.3 cu m/ha in Guangxi provenance, although the better performing provenіences reached 26.6 cu m/ha in 4.4 years.

7.3 The sale price of well grown Eucalyptus tree after 7 years is about Rs. 100-150/per bole. The sale price of wood at the farm gate varies from Rs. 800-1000/metric tonne.

Commercial uses of Eucalyptus

- Eucalyptus is one of the fastest growing trees and is an excellent timber for paper and pulp, particleboard and hardboard industries.

- It is also an excellent source of fuelwood and charcoal.
- Eucalyptus wood is also used for light and heavy construction, railway sleepers, bridges, piles, poles and mining timber.
- Indian Standards are available for use of *E. tereticornis* timber, after treatment, for door frames, window shutters, furniture, cabinet, tool handles, packing cases and crates.
- Leaf extracts of the species have pesticidal properties and can be promoted as a biopesticide.
- The leaves of the species are rich in essential oils, that have many medicinal uses. *Eucalyptus globulus* can be raised commercially for Eucalyptus oil.
- *E. tereticornis* is a major source of pollen in apiculture and produces a medium amber honey of distinctive flavour.
- The wood and bark of the tree have a tannin content of 6-12% and 3-15% respectively, though not used as a commercial source of tannin.
- Eucalyptus is a large ornamental tree suitable for parks and avenue plantations.
- The tree may be used as an agro-forestry species. Eucalyptus in combination with pineapple have given excellent results in China.
- The tree species can be effectively used for regeneration of denuded lands and prevention of soil erosions in drought -affected areas.

Unit cost and Economics

Unit cost for raising Eucalyptus under high density plantations (1.5m x 1.5m) has been worked out to Rs. 50,800/- per hectare. Considering harvesting at the age of 7 years, with a sale price of Rs. 80/pole and fuelwood @ Rs. 500/MT, the IRR works out to 36%. The details of techno-economic parameters and economics are furnished below. The investment has been found to be technically feasible, financially viable and bankable.

Unit Cost for raising Eucalyptus hybrid in 1 ha of wasteland

Spacing: 1.5m x 1.5m.
No. of plants per ha. - 4444
Mortality replacement: 20%

Avg. wage rate Rs. 66 per Man Day
Interest on term loan 12%
Margin: 10% of the unit cost

S. No.	Particulars of Works	Unit	Cost (Rs.) per year			Total (Rs.)
			1	2	3	
1.	Site preparation, ploughing, alignment and staking	Lump Sum (LS)	1000			1000
2.	Digging of pits (45cm x 45cm x45cm) and refilling with soil, fertilizers, Organic Manure and insecticides	@ Rs. 4 per pit	17776	3552	-	21328
3.	Cost of planting material including transportation 4444/888 plants	@ Rs. 2/- per plant	8888	1776	-	10664
4.	Cost of planting and replanting	44md/9 MD	2904	594	-	3498
5.	Cost of Aldex/BHC - anti-termite pesticide (Benzene Hexa Chloride powder)	LS	150			150
6.	Cost of NPK @ 25 gms/plant	@ Rs.8/kg	889	889	889	2667
7.	Cost of Rock Phosphate @ 50 grms /plant	@ Rs. 2.20/kg	489			489
8.	Cost of irrigation -during I and II year	@ Rs. 1000/ha	1000	1000	-	2000
9.	Weeding	LS	1000	1000	-	2000
10.	Earth working including application of fertilizers (3,2,2)	10md /Working	1980	1320	1320	4620
	Sub Total		36076	10131	2209	48416
	Contingency @ 5%		1804	507	110	2421
	Grand Total (Unit cost taking first 3 yrs of expenditure)		37880	10638	2319	50837
		round off	37900	10600	2300	50800
	Unit cost per ha.					50800
	Margin @10%		3790	1060	230	5080
	Bank loan @ 90%		34110	9540	2070	45720

Other expenditure

Annual Maintenance cost of Rs. 1000/year towards fertilization/manuring. Harvesting cost in 7th and 14 th years : Rs. 10,000 /- and 8,000/- respectively. Pruning (of coppice) in the 8th year: Rs. 1000/-

Yield and Income per hectare

Year	No. of Trees	Sale price of timber Rs/pole	Income	Fuelwood generated at harvesting (MT)	Sale price of fuelwood (Rs/MT)	Income from fuelwood	Total income
7	3555	80	284,400	20	500	10000	294,400
14	2844	80	227,520	20	500	10000	237,520

Income and expenditure Statement

Particulars	Year 1	Year 2	Year 3	Year 4	Year 5	Year 6	Year 7	Year 8	Year 9	Year 10	Year 11	Year 12	Year 13	Year 14
Capital cost	37,900	10,600	2,300	-	-	-	-	-	-	-	-	-	-	-
Recurring cost	-	-	-	1000	1000	1000	10,000	2000	1000	1000	1000	1000	1000	8000
Total Expenditure	37,900	10,600	2300	1000	1000	1000	10,000	2000	1000	1000	1000	1000	1000	8000
Income	-	-	-	-	-	-	294,400	-	-	-	-	-	-	237520
Net Income	-37900	-10600	-2300	-1000	-1000	-1000	284,400	-2000	-1000	-1000	-1000	-1000	-1000	229520

NPV of costs : 50625 NPV of benefits :144244 NPV: 93619 CR: 2.85 : 1 IRR: 36.37%

Repayment

Particulars / Years	1	2	3	4	5	6	7	---------	14 yr.
Loan outstanding at the beginning of the year	34110	43650	45720	45720	45720	45720	45720		
Interest @12%	4093	5238	5486	5486	5486	5486	5486		
Cumulative interest	4093	9331	14817	20303	25789	31275	36761		
Net Income	0	0	0	0	0	0	284400		229520
Repayment towards interest							36761		
Repayment towards principal							45720		
Total repayment							82481		
Net surplus	**0**	**0**	**0**	**0**	**0**	**0**	**201919**		**229520**
Loan Outstanding at the end of the year	34110	43650	45720	45720	45720	45720	0		0
Interest outstanding at the end of the year	4093	9331	14817	20303	25789	31275	0		0

Repayment period: 7 years with 6 years moratorium (freeze)

Forward and Backward Linkages

The planting stock and technical guidance for commercial plantations of Eucalyptus is available from the Forest Departments, Forest Research Institutes, Agricultural Universities/Research Institutes and Wood-Based User Industries.

Eucalyptus is one of the most popular raw materials for paper and pulp, particleboard and hardboard industries and also an excellent source of fuelwood and charcoal. Many wood-based industries have tie-up arrangement with farmers/entrepreneurs for raising Eucalyptus plantation for their consumption. Market for the species is, thus, not a problem, if cultivated with proper tie-ups. List of important wood based industries in the country is listed in Annexure.

Repayment

- Repayment of the bank loan can be made in 7 years with 6 years' moratorium.
- Rate of Interest, Margin and Security:
- With deregulation of interest rates, the rate of interest charged to the ultimate borrower is decided by the financing banks. Margin money and Security may be charged by the bank as per the instructions issued by the RBI/NABARD from time to time.

NABARD Refinance

In tune with the National priorities, NABARD extends refinance support for promoting wasteland development/agroforestry through Eucalyptus cultivation at a concessional rate of interest.

Exercise-3

3. Study of Computations of Depreciation

Depreciation : The decline in value of capital equipment due to wear and tear is called as depreciation. It is caused by two factors *viz., time and use.* As depreciation continues, the serviceability and value of the asset diminishes.

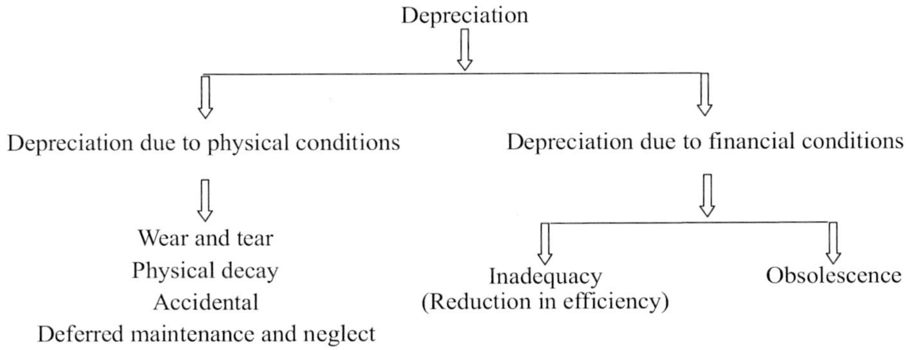

Methods of computation of depreciation

1. Straight line method
2. Annual revaluation method
3. Diminishing balance method
4. Sum of the year-Digits method or Reducing Fraction method
5. Compound interest methods
   a) Sinking fund method
   b) Annuity charging method
6. Insurance policy method
7. Machine hour basis method

1. Straight line method

This method is easy, simple and usually very satisfactory for most purposes. This method assumes that assets are used more or less to the same extent every year and therefore, equal amounts of costs on account of their use can be charged every year.

$$\text{Annual Depreciation} = \frac{\text{Original cost - Junk value}}{\text{Expected life of the asset (No. of useful years of life)}}$$

Junk value otherwise also called as scrap value, salvage value or residual value. The purchase price minus junk value is also known as total anticipated depreciation.

Example 1. A bullock cart costs Rs. 5000 and is expected to last for 10 years; the salvage value after 10 years is Rs. 500. Then calculate the annual depreciation value with the help of the following information in the table given below :

Table: Example for Straight line method

Life period	Value at the beginning of the year (Rs.)	Amount to be depreciated (Rs.)	Depreciated value at the end of the year (Rs.)
1.	5000	450	4550
2.			
3.			
4.			
5			
6.			
7.			
8.			
9.			
10.			

2. Annual Revaluation Method

This method estimates the market value of the asset in the beginning and the end of the year and then taking the difference as depreciation. This method is, however, useful in case of livestock in the early years of life, i.e. in appreciating phase. For items not bought and sold frequently, it becomes difficult.

3. Diminishing Balance Method

A fixed rate of depreciation is used for every year and applied to the value of the asset at the beginning of the year. The original cost of an asset is divided by its estimated life to knock off a fixed percentage. This percentage is deducted every year from the diminished balance, till the asset reached the salvage value and no further depreciation is possible.

$$\text{Therefore, annual depreciation} = \frac{\text{Original cost}}{\text{Expected life}}$$

i.e. The amount of depreciation is deducted every year from the diminished balance value of the asset, till it reaches its salvage value. The amount of depreciation is different at different stages of the asset and gradually diminishes with the life.

Example: Suppose a chain saw is purchased for Rs. 1000 and its expected life is 10 years. Then calculate the annual depreciation.

Years of life	Value at the beginning of the year (Rs.)	Amount to be depreciated (10% of diminished value in Rs.)	Depreciated value at the end of the year (Diminished balance value in Rs.)
1.	1000	100	900
2.			
3.			
4.			
5.			
6.			
7.			
8.			
9.			
10.			

4. Sum of the year – Digits method (Reducing Fraction method)

The annual depreciation is found out by multiplying a fraction times the amount to be depreciated (cost minus the salvage value). We can determine the fraction for any year by the following formula:

$$\text{Fraction for any year} = \frac{\text{The years of life remaining at the beginning of accounting period}}{\text{The sum of the years of life of the assets}}$$

$$\text{Fraction for first year} = \frac{10}{1+2+3+4+5+6+7+8+9+10} = \frac{10}{55}$$

$$\text{Fraction for second year} = \frac{9}{1+2+3+4+5+6+7+8+9+10} = \frac{9}{55}$$

Just as we compute depreciation by straight line method, similarly we deduct salvage value from original cost each year for finding the annual depreciation.

Rate of annual depreciation = (Original cost - Junk value) x Fraction for the particular year

Example: Suppose the value of a tractor is Rs. 92,000 and expected to last for 10 years. Then calculate the annual depreciation.

Year	Value at the beginning of year	Annual depreciation	Remaining balance
1.			
2.			
3.			
4.			
5.			
6.			
7.			
8.			
9.			
10.			

5. Compound Interest Method

a) Sinking Fund Method

In this system, a depreciation fund equal to the actual loss in the value of the asset or machine is estimated, taking into account, the interest on the so accumulated fund. The rate of depreciation will be constant throughout the life of the machine.

Let D = Rate of depreciation per year

R = Rate of interest on accumulated fund in fraction number

C = Total cost of machine

S = Scrap value

N = No. of years of life of machine

$$\text{Therefore, } D = \frac{R(C-S)}{(1+R)^N-1}$$

Example 1. A machine is purchased for Rs. 40,000. The estimated life of machine is 15 years and scrap value Rs. 15,000. If the rate of interest on depreciation fund is charged at 5%, then calculate the rate of depreciation by sinking fund method.

Example 2. A mechanized farm is having a sugarcane harvest with initial value of Rs. 2,00,000 and the salvage value of Rs. 20,000 at the end of 20 years but is sold for Rs. 1,45,000 at the end 10 years. Calculate the profit or loss, if sinking fund depreciation method at 8% compounded annually, was adopted.

b) Annuity Charging Method

In this method, interest is charged on the cost of machine or assets every year on the book value, but the rate of depreciation is constant every year.

Let C = Cost of machine

S = Scrap value

N = No. of years of machine life

R = Rate of interest in fractions

D = Rate of depreciation

If the value of machine after 1 year becomes C_1, then,

$D = CR + C - C_1 = C (1 + R) - C_1$

In the same way, the value of machine after 2 years will be say $C_{2,}$ then

$D = C_1 R + C_1 - C_2 = C_1 (1 + R) - C_2$

Hence, the standard formula will be:

$$D = \frac{\{C (1 + R)N - S\} \{1 - (1 + R)\}}{1-(1+R)^N}$$

Example 1. Find the depreciation annuity by the annuity charging method after 3 years, when the cost of pumpset is Rs. 8000 and scrap value is Rs. 4000. Rate of interest is 5%.

6. The Insurance Policy Method

This method covers the risk, if the machine becomes unserviceable before its estimated life. In this method the machine is insured with the insurance company and premiums are paid on the insurance policy. When the policy matures, the company provides sufficient sum to replace the machine.

7. Machine Hour Basis Method

In this method, the rate of depreciation is calculated, considering the total number of hours a machine runs in year, and therefore, a work-hour chart of every machine is maintained to know the total number of hours the machine runs in a year.

Example 1. A machine is costing Rs. 11,000 and expected to run for 10 years, at the end of which its scrap value is likely to be R. 1000. The machine is expected to run 2000 hours/year on the average. Estimate the depreciation charges per hour of the machine.

Exercise-4

4. Study of Analysis of Net Worth Statement of Forestry Enterprise(Balance Sheet)

Net Worth Statement

The net worth statement shows the financial condition and stability of the business at a particular point of time. Net Worth is shown as an excess of assets over liabilities, i.e. the liability of the business to the farmer or farmer's claim in the business.

Liabilities

The debts or amounts of money owned by an individual, partnership or corporation to others are called liabilities. Liabilities are the commitments of the farmer. This may be in the form of loans, promissory notes, material bought on credit, etc.

Liabilities are of three types :

1. **Long duration liabilities**: These liabilities are those which do not require repayment during current accounting period, e.g. long term loans.
2. **Intermediate liabilities**: These liabilities can be differed or postponed for the present, but fall due within the year, e.g. promissory notes, medium term and short term loans.
3. **Current liabilities**: Repayment of these liabilities may be immediately paid.

In summary form, the balance sheet can be illustrated as follows:

Liabilities = Net capital =__________	Assets =
On the left are the liabilities, that is, what the business owes.	On the right han d side are assets of the business that is what the business is worth.

The net capital (also known as capital balance, net worth or equity) is the balancing item.
Thus Net capital = assets – liabilities

Net worth or equity is the sum of what would remain to the owner, if the business were sold and all the liabilities paid. If there were a net capital deficit instead of a surplus, this would have to be included on the assets side to balance, and it would mean that the business was insolvent. The details of the asset side of the balance sheet show how the capital is being used in the business, and those on the liabilities side the sources of that capital.

Net worth statement is thus prepared for the farmer and not for his business. It shows whether the business is expanding or shrinking. The business is said to be solvent, when the net worth or equity is greater than zero.

Analysis of Net Worth Statement

The net worth calculated above shows only the absolute amount, by which total assets differ from total liabilities at a point of time. This may not give the true picture of the financial position of the farmer :

1. Net capital Ratio
2. Working Ratio
3. Current Ratio

1. **Net capital Ratio :** Net Capital Ratio is the ratio of total assets and total liabilities.

$$\text{Net Capital Ratio} = \frac{\text{Total assets}}{\text{Total liabilities}}$$

It is measured to work out the degree of financial safety over a period of time. Higher the ratio, the safer is the business and the less vulnerable to an unexpected drop in the value of its assets or losses.

Comparison of Net worth for two farmers

Net worth for two farmers may be the same, but net capital ratio would be different.

Farmer (A)	Farmer (B)

Conclusion

Net worth statement of a farmer as on 1st January, 2014 is as follows:

Liabilities	Value (Rs)	Assets	Value (Rs)
1. Current liabilities		**1. Current assets**	
a. Short term loans		a. Cash in bank	
b. Hand loans		b. Cash on hand	
c. Fertilizers		c. A/c receivable	
d. Feeds		d. Feed, grain etc.	
Sub total		**Sub total**	
2. Intermediate liabilities		**2. Working assets**	
Loans on machinery and equipment		Machinery and equipment	
Loans on milch animals		Livestock	
Sub total			
3. Fixed liabilities		**3. Fixed assets**	
		Land	
		Farm buildings	
Sub total		**Sub total**	
Total of all liabilities		**Total of all assets**	
Net worth			

Net worth as on 01.01.2014 = Total assets - Total liabilities

2. **Working Ratio :** This measures the financial safety of the business over an intermediate period of time.

$$\text{Working Ratio} = \frac{\text{Sum of working and current assets}}{\text{Sum of medium term liabilities and current liabilities}}$$

3. **Current Ratio :** Current Ratio is the ratio of current assets and current liabilities

$$\text{Current Ratio} = \frac{\text{Current assets}}{\text{Current liabilities}}$$

It measures the degree of immediate solvency. The aspect of stability described in net capital ratio relates to the long term solvency of the business. However, the business also has to survive in the shorter run, weathering the temporary stresses and strains that may occur. An important measure in this regard is the current ratio. The higher the ratio, the safer is the firm in the short run, because the more likely it is to survive unexpected demands from creditors.

$$\text{Current Ratio} = \frac{\text{Current assets}}{\text{Current liabilities}}$$

All these ratios can be compared overtime. Wider these ratios, the better will be the financial position of the business.

References

Dexter, K. and D. Barber (1960). *Farming for Profits*, Penguin, Harmondsworth.

Dhondyal, S.P. (1985), Farm Management, Friends Publication, Meerut (India).

Dillon, J.L. (1980). 'The Definition of Farm Management', *Journal of Agricultural Economics* 31: 257-8.

Doll, John P. and Orazem. F. (1984), Production Economics: Theory with Application, John Wiley and Sons, New York. Castle, E.N., M.H. Becker and A.G. Nelson (1987). *Farm Business Management*, 3rd edn, Macmillan, New York.

Doll, John P. and Orazem. F. (1984), Production Economics: Theory with Application, John Wiley and Sons, New York.

Giles, T. and M. Stansfield (1995). *The Farmer as Manager*, 2nd edn, CAB International, Wallingford.

Heady, E.O. and H.R. Jensen (1954). *Farm Management Economics*, Prentice-Hall, Englewood Cliffs.

Hellriegel, D. and J.W. Slocum (1986). *Management*, 4th edn, Addison-Wesley, Reading.

Hitt, M.A., R.D. Middlemist and R.L. Mathis (1989). *Management: Concepts and Effective Practice*, 3rd edn, West Publishing Co., Saint Paul.

Jensen, H.R. (1977). 'Farm Management and Production Economics, 1946-70', *A Survey of Agricultural Economics Literature*, University of Minnesota Press, Minneapolis, Vol. 1, pp. 1-89.

Johl, S.S. and Kapoor, T.R. (1973), Fundamentals of Farm Business Management, Kalyani Publishers, Ludhiana.

Kahlon, A.S. and Karam Singh (1992), Economics of Farm Management, Allied Publishers, New Delhi.

Kast, R.E. and J.E. Rosenzweig (eds) (1974). *Organization and Management: A Systems Approach*, 2nd edn, McGraw-Hill Kogakusha, Tokyo.

Kay, R.D. and W.M. Edwards (1994). *Farm Management*, 3rd edn, McGraw-Hill, New York.

Koontz, H. (1969). 'A Model for Analyzing the Universality and Transferability of Management', *Academy of Management Journal* 12:415-30.

Koontz, H. (1976). 'The Management Theory Jungle', in H. Koontz and C. O'Donnell (eds), *Management: A Book of Readings*, McGraw-Hill, New York, Ch. 3.

Mintzberg, H. (1983). *Structure in Fives: Designing Effective Organizations*, Prentice-Hall, Englewood Cliffs.

Nix, J. (1979). 'Farm Management: The State of the An (or Science)', *Journal of Agricultural Economics* 30: 277-91.

Rae, A.N. (1994). *Agricultural Management Economics: Activity Analysis and Decision Making*, CAB International, Wallingford.

Raju, V.T. and Rao, D.V.S. (1990), Economics of Farm Production and Management, Oxford & IBH Publishing Co. Pvt. Ltd., New Delhi-110 001.

S.S. Johl and T.R. Kapur. Fundamentals of Farm Business Management

Sankhayan, P.L. (1988), Introduction to the Economics of Agricultural Production, Prentice Hall of India Private Limited, New Delhi-110 001.

Sankhayan, P.L. (1988), Introduction to the Economics of Agricultural Production, Prentice Hall of India Private Limited, New Delhi-110 001.

Schultz, T.W. (1939). 'Theory of the Firm and Farm Management Research', *Journal of Farm Economics* 21: 570-86.

V.T. Raju and D.V.S. Rao. Economics of Farm Production and Management.

❑❑❑

Chapter 15

Model Questions Bank

Q. Choose the Correct Answer for the Following

1. The net revenue obtained when the marginal cost (MC) is equal to the price of the product (MR) is:
   a) Minimum b) Maximum
   c) Neutral d) Variable
2. It is worthwhile to invest on the project, if the Benefit Cost Ratio is:
   a) More than one b) Positive
   c) Negative d) None
3. This production relationship is explained by the principle of product substitution and law of equimarginal returns:
   a) Factor-Factor b) Product-Product
   c) Factor-Product d) None
4. The loss of value of an asset over a period of time is called as:
   a) Depreciation b) Deflation
   c) Appreciation d) None
5. The basic measure of efficiency in the use of capital employed by the business is:
   a) Net Present Value b) Rate of capital turn over
   c) Benefit Cost Ratio d) None of these
6. If the services of some resources (e.g. labour, building, etc.) are not used, those cannot be stocked and such services are known as:
   a) Variable resources b) Stock resources
   c) Fixed resources d) Flow resources

7. Agricultural prices and production usually move in ____________ directions.

   a) Opposite b) Horizontal

   c) Equal d) Opposite or equal

8. This enterprise relationship exists when increase or decrease in the level of one product does not affect the production level of the other product:

   a) Competitive b) Supplementary

   c) Complementary d) Antagonistic

9. Algebraically, this relationship can be expressed as Y1=f (Y2 Y3, ... Yn)

   a) Factor - Factor b) Factor - Product

   c) Product - Product d) None

10. The demand for agricultural products (Eg. food) is:

    a) Inelastic b) Constant

    c) Elastic d) None of these

11. This production relationship is explained by the principle of factor substitution or principle of substitution between inputs:

    a) Factor-Factor b) Product-Product

    c) Factor-Product d) None

12. The Line which defines all the possible combinations of two commodities which would yield an equal revenue or income:

    a) Iso quant curve b) Production Possibility Curve

    c) Iso revenue curve d) None

13. Most of the small businesses can be classified as the following type/s:

    a) Retailing b) Franchising

    c) Both a & b d) None of these

14. This curve represents all possible combinations of two resources physically capable of producing the same quantity of output:

    a) Iso quant curve b) Isoproduct curve

    c) Both a & b d) None

15. The turnover rate in agriculture or forestry is relatively __________ because the production process requires a period of several years.

a) Slow b) Constant

c) Fast d) Variable

16. The resources which are not used in one period of production can be stored for a later period are known as:

a) Variable resources b) Stock resources

c) Fixed resources d) Flow resources

17. The Assets which are most liquid and are consumable in a year are called as:

a) Working assets b) Fixed assets

c) Current assets d) None

18. A farmer performs the following function/s himself:

a) Farm manager b) Farm executive

c) Farm entrepreneur d) All of these

19. The ratio of current assets and current liabilities is known as:

a) Current ratio b) Net capital ratio

c) Working ratio d) None

20. The loan made for a short period from a few days to several months is known as:

a) Long term loan b) Short term loan

c) Both a & b d) None

21. The determination of least cost combination of resources is concerned with this production relationship:

a) Factor-Product b) Product-Product

c) Factor-Factor d) None

22. Management of forests broadly involves the following main task/s:

a) Harvesting b) Marketing

c) Both a & b d) None of these

23. The goal of Product-Product relationship is:
    a) Profit maximization   b) Optimization of production
    c) Cost minimization   d) None

24. The forest efficiency measures can be categorized into:
    a) Aggregate measures   b) Ratio measures
    c) Both a & b   d) None

25. Land use efficiency is grouped under this forest efficiency measures:
    a) Aggregate measures   b) Ratio measures
    c) Absolute measures   d) None

26. Gross returns minus all costs is equal to:
    a) Net returns   b) Capital returns
    c) Total cost   d) None

27. Asia's first stock exchange established in 1875 is ___________
    a) Toronto Stock Exchange
    b) Kualalumpur Stock Exchange
    c) The Bombay Stock Exchange
    d) None

28. The largest Stock Exchange in the World's Stock Market is ___________________
    a) New York Stock Exchange   b) NASDAQ Stock Exchange
    c) London Stock Exchange   d) None

29. The Stock Market for mainly technology shares is ____________
    a) Toronto Stock Exchange   b) NASDAQ Stock Exchange
    c) New York Stock Exchange   d) None

30. An example for the type of stock exchange of a virtual kind composed of a network of computers, where trades are made electronically via traders' is______________
    a) NASDAQ Stock Exchange   b) Amsterdam Stock Exchange
    c) London Stock Exchange   d) None

31. An individual or institution (including a corporation) that legally owns a share of stock in a public or private corporation is called as______________

a) A shareholder b) A stockholder
c) Both a & b d) None

32. The aggregation of buyers and sellers of stocks or shares which is a loose network of economic transactions but not a physical facility or discrete entity is called as______________

a) Stock Market b) Equity market
c) Both a & b d) None

33. This is a form of business in which two or more people operate for the common goal of making profit:

a) Partnership b) Sole proprietorship
c) Corporation d) None

34. A privately owned business owned by one person who may operate on his or her own or may employ others is known as:________________

a) Partnership b) Sole proprietorship
c) Corporation d) None

35. A __________________is privately held or publicly traded, owned by multiple shareholders which is overseen by a board of directors and hires the business's managerial staff:

a) Partnership b) Sole proprietorship
c) Corporation d) None

36. This stock exchange is the largest electronic screen-based equity securities trading market in the United States:

a) Toronto Stock Exchange b) NASDAQ Stock Exchange
c) New York Stock Exchange d) None

37. A tractor costs Rs.24,000 when new and expected to last 10 years and the junk value of the tractor after 10 years would be Rs.4,000. Then annual depreciation of a tractor would be _______.

a) Rs.1000 /year b) Rs.2000 /year
c) Rs.3000 /year d) Rs.4000 /year

38. _______________is a form of selective distribution in case of a small business.

a) Personal services b) Retailing

c) Professional services d) Franchising

39. Management of forests broadly involves the following main task/s:

a) Control of composition and structure of the growing stock

b) Harvesting and marketing of forest produce

c) Administration of forest property and personnel

d) All of these

40. Algebraically, this relationship can be expressed as Y = f (X1 / X 2,X3............Xn):

a) Factor - Product b) Factor - Factor

c) Product - Product d) None

41. This principle explains the product-product relationship and helps in deciding the optimum combination of products:

a) Principle of substitution between inputs

b) Principle of equi-marginal returns

c) Principle of substitution between products

d) Principle of variable proportions

42. Of the total costs, portion of fixed costs is ___________ in agriculture than industry.

a) Less b) Meutral

c) Variable d) More

43. This curve is a convenient method for compressing three dimensional picture of production into two dimensions:

a) Iso quant curve b) Production Possibility Curve

c) Iso revenue curve d) None

44. This curve is known as Opportunity Curve because it represents all production possibilities or opportunities available with limited resources:

a) Iso quant curve b) Production Possibility Curve

c) Iso revenue curve d) None

45. This curve is called as Iso resource Curve or Iso factor curve because each output combination on this curve has the same resource requirement:
    a) Iso quant curve
    b) Iso revenue curve
    c) Production Possibility Curve
    d) None
46. This curve is called as Transformation curve, as it indicates the rate of transformation of one product into another:
    a) Production Possibility Curve
    b) Iso quant curve
    c) Iso revenue curve
    d) None
47. The following enterprises compete for farm resources and substitute for each other:
    a) Complementary enterprises
    b) Competitive enterprises
    c) Supplementary enterprises
    d) Antagonistic products
48. This relationship holds good when increase or decrease in the production of one product affects the production of the other commodity inversely:
    a) Complementary enterprises
    b) Antagonistic products
    c) Supplementary enterprises
    d) Competitive enterprises
49. In this following relationship, the products are produced through a single production process and one of the products cannot be produced alone but must be accompanied by one or more of other products:
    a) Complementary enterprises
    b) Antagonistic products
    c) Supplementary enterprises
    d) Joint Product enterprises

50. This relationship holds good when two products may be detrimental to each other because of disease or similar factors where only one of the products should be produced:

a) Antagonistic products b) Complementary enterprises

c) Supplementary enterprises d) Competitive enterprises

51. Two products are said to be in this following relationship, when an increase in output of one results in an increase in output of the other, with resource amount held constant:

a) Complementary enterprises

b) Antagonistic products

c) Supplementary enterprises

d) Competitive enterprises

52. This relationship exists when increase or decrease in the level of one product does not affect the production level of the other product:

a) Complementary enterprises

b) Antagonistic products

c) Supplementary enterprises

d) Joint Product enterprises

53. Antagonism relationship is the opposite of the following relationship:

a) Competitive enterprises b) Antagonistic products

c) Supplementary enterprises d) Complementary enterprises

54. Laws of returns or Principle of variable proportion in forest business management consists of the following phase/s:

a) Diminishing return b) Constant returns

c) Increasing returns d) All of these

55. The following law can be stated as "If increasing amounts of one input are added to a production process while all other inputs are held constant, the amount of output added per unit of variable input will eventually start decreasing":

a) Diminishing returns b) Constant returns

c) Increasing returns d) None

56. The following principle of forest business management states that "A limited input should be allocated among alternative uses in such a way that the marginal value products of the last unit are equal in all its uses":

a) Principle of substitution between inputs

b) Principle of equi-marginal returns

c) Principle of substitution between products

d) Principle of variable proportions

57. According the Principle of Costs, the maximum net revenue is obtained when:

a) Marginal cost (MC) is more than the price of the product (MR)

b) Marginal cost (MC) is less than the price of the product (MR)

c) Marginal cost (MC) is equal to the price of the product (MR)

d) Marginal cost (MC) is constant

58. Algebraically, this relationship can be expressed as Y = f (X1 X2 / X3 X4 …….. Xn):

a) Factor – Product b) Factor – Factor

c) Product – Product d) None

59. This economic principle explains one of the basic production relationships *viz.*, F-F relationship guiding in the determination of least cost combination of resources:

a) Principle of substitution between inputs

b) Principle of equi-marginal returns

c) Principle of substitution between products

d) Principle of variable proportions

60. As India has distinct seasons of production, the agricultural year (period) which may be more suitable and convenient "*accounting period*" for most parts of our country is:

a) Starting from 1st January and ending 30th December

b) Starting from 1st July and ending 30th June of next calendar year

c) Starting from 1st February and ending 30th January of next calendar year

d) Starting from 1st April and ending 30th March of next calendar year

61. It is a method of recording each transaction in the books of accounts in its two-fold aspects, i.e. two entries are made for each transaction in the same set of books, one being a debit entry and the other a credit entry:

a) Double entry system b) Single entry system

c) Both a & b d) None

62. In case of decreasing returns, the production curve is:

a) Convex to X axis b) Concave to X axis

c) Parallel to Y axis d) Parallel to X axis

63. If two goods are supplements, this means that a rise in the price of one commodity will induce:

a) An upward shift in demand for other commodity

b) A rise in the price of the other commodity

c) A downward shift in the demand for the other commodity

d) No shift in the demand for the other commodity

64. The art of managing a forest successfully is measured by:

a) Increase in forest area

b) Maintenance of scarce resources

c) Extension

d) Profitableness

65. These decisions involve less investment and are made more frequently:

a) Operational b) Administrative

c) Marketing d) None of the above

66. Production Economics is concerned with:

a) Choice of production patterns

b) Resource use

c) Maximization of profit

d) All of the above

67. The amount of output with given amounts of factors of production is:

a) Technical Efficiency b) Economic Efficiency

c) Economic system d) Optimum Efficiency

68. Factor - Factor relationship guides the producer in deciding:

a) How much to produce b) What to produce

c) How to produce d) When to produce

69. At the point of inflection elasticity of production:

a) Minimum b) Less than 1

c) Greater than 1 d) Constant

70. Selection of enterprise is:

a) Strategic management decision

b) Operational

c) Administrative

d) Marketing

71. At inflection point on TP curve:

a) AP is maximum b) MP is zero

c) MP is negative d) MP is maximum

72. Factor-Factor relationship deals with:

a) Iso-cost b) Iso-product

c) Iso-quant d) All the above

73. MRS = 0 at:

a) Ridge line b) Isoclines

c) Expansion path d) Budget line

74. In complimentary relationship Production Possibility Curve is:

a) Upward sloping b) Downward sloping

c) Straight line d) Convex to origin

75. Which among the following is highly perishable?

a) Land b) Labour

c) Capital d) Organization

Q. Match the Following

Set I

Part A	Part B
1. Facilitative service	a) Phase of distribution
2. Bee-keeping	b) Process of transformation
3. Retailing	c) Resources are not fixed
4. Franchising	d) Integrated Science
5. Short run production function	e) Consultancy and insurance
6. Farm management science	f) Agro service centre
7. Production	g) Diet service, fast food, etc.
8. Personal services	h) Resources are fixed
9. Farm Management Decisions	i) Apiculture
10. Long run production function	j) Business Principles and Practices

Set II

Part A	Part B
1. Facilitative service	a) Phase of distribution
2. Bee-keeping	b) Process of transformation
3. Retailing	c) Resources are not fixed
4. Franchising	d) Integrated Science
5. Short run production function	e) Consultancy and insurance
6. Farm management science	f) Agro service centre
7. Production	g) Diet service, fast food, etc.
8. Personal services	h) Resources are fixed
9. Farm Management Decisions	i) Apiculture
10. Long run production function	j) Business Principles and Practices

Set III

Part A	Part B
1. Isoquants	a) Phase of distribution
2. Long run production function	b) Iso-resource Curve
3. Retailing	c) Resources are not fixed
4. Franchising	d) Iso -product curves
5. Short run production function	e) Consultancy and insurance
6. Basic measure of efficiency	f) Agro service centre
7. Production Possibility Curve	g) Diet service, fast food, etc.
8. Personal services	h) Resources are fixed
9. Facilitative service	i) Apiculture
10. Beekeeping	j) Rate of capital turn over

Set IV

Part A	Part B
1. Basic measure of efficiency	a) Sandal tree species with host tree sps
2. Long run production function	b) Iso-resource Curve
3. Production	c) Resources are variable
4. Facilitative service	d) Iso -product curves
5. Short run production function	e) Consultancy and insurance
6. Isoquants	f) *Ceiba pentandra* (silk cotton) lint and seeds
7. Production Possibility Curve	g) Aquaculture and paddy cultivation
8. Joint Product enterprises	h) Resources are fixed
9. Complementary enterprises	i) A process of transformation
10. Antagonistic products	j) Rate of capital turn over

Set V

Part A	Part B
1. Iso-quants	a) Selective distribution of goods
2. Depreciation	b) Iso-resource Curve
3. Current assets	c) Debit and Credit
4. Franchising	d) Iso -product curves
5. Production function	e) Liquid assets
6. Double entry system	f) Junk value
7. Production Possibility Curve	g) A business owned by one person
8. Corporation	h) Law of Diminishing returns.
9. Sole proprietorship	i) Multiple shareholders
10. Factor-Product relationship	j) Relation between inputs and outputs

Set V

Part A	Part B
Tariff	a) Cords or board feet
Liquid assets	b) Payment at a fixed interval
Liabilities	c) Taxes on goods transported
Modern economy	d) Lump sum estimate
Forest insurance	e) Cash, bank deposits
Income	f) Overhead cost
Consumption tax	g) Wages, materials
Stumpage value	h) Money economy
Royalty sale	i) Production of timber
The timber cruise	j) Government grants and tax rebates

Set VI

Part A	Part B
Tangible benefits	a) Service sector
Liquid assets	b) Shift in supply curve
Perfectly elastic demand	c) Shift in demand curve
Intangible benefits	d) Cords or board feet
Project feasibility	e) Agriculture and allied sector
Primary sector of economy	f) Benefit cost ratio
Input demand	g) Ecosystem services of forest
Change in income technology	h) Coefficient is infinity
Change production technology	i) Cash, Bank deposits
Tertiary sector of economy	j) Timber products

Q. State True or False

1. The production function portrays an input –output relationship.
2. All products are produced with single input.
3. In the long run all costs are considered variable costs.
4. Insurance other than life insurance may consider one of the overhead costs of business.
5. Tariffs are the taxes on goods transported from one country to another.
6. The process determining the monetary value of timber is a forest appraisal.
7. Ridge line denotes the limit of substitution.
8. The working assets are most liquid than fixed assets, such as farm machinery and equipments, etc.
9. Depreciation represents the loss in value of a machine over its working life.
10. Declining value method of depreciation is known as reducing balance method.
11. Economics is concerned with behavior in the activities of production, consumption and exchange of goods and services.
12. The advance payment of sums due at later dates is credit.
13. The land, timber, other vegetative and physical improvement is forest capital.

14. Long duration liabilities are those, which do not require repayment during the current accounting period.
15. The enterprise combination is often called product-product relationship.
16. Forest economics is an applied field of economics, which deals more with problems in application than with theory.
17. The production function portrays an input-output relationship.
18. All products are produced with a single input.
19. Ability and willingness to purchase in given point of time is known as demand.
20. Rising trend of prices in the economy is a common phenomenon.
21. In the long run, all costs are considered variable costs because all inputs are variable.
22. Insurance other than life insurance may be considered one of the overhead costs of business.
23. In a producers co-operative, the more a member sells through it, the smaller his profit.
24. The expression stumpage value often is loosely used.
25. Long term loan are made for long period.
26. In 2nd stage of classical production function Ep > 1
27. Optimum output criteria in Factor-Product relationship is MP = Ap
28. Forest management is an integrated science
29. An operation management decision involves very high investments
30. At least cost combination of two inputs iso-cost line is tangent to iso resource curve
31. Fixed costs vary with the level of production
32. Farm management concerned with rational use of resources from the individual farmer point of view.
33. Two enterprises are said to be supplementary, when an increase in one enterprise will add to increase in the production of other enterprise.
34. Enterprise budgeting enterprise estimates costs and returns on minor modifications in the farming practices instead for the farm business as a whole.
35. At the point of least cost combination of inputs, the ratio between MRTS between inputs and inverse price ratio will be unity.

Q. Fill In the Blanks with Suitable Answers

1. ________________ is the cost of producing additional unit output.
2. ____________________ means ability and quantity offered for scale in given point of time.
3. ______________, ______________& ______________are said to be factors of production.
4. Taxes which are collected not on what the tax-payer owns but on his income, net or gross, are called as ______________.
5. ______________ are taxes on goods transported from one country to another.
6. ______________ is a loan made upon the security of a piece of real estate.
7. ______________ refers to that part of man's wealth which is used in producing further wealth.
8. ______________ are lines or curves that pass through points of equal marginal rates of substitution.
9. ________________ plays a very important role in deciding any business activity.
10. Ratio of current assets and current liabilities is called as ____________________
11. Gross return minus all cost is equal to ________________
12. The goal of Product–Product relationship is ______________
13. A business organization owned by a single person is known as __________________
14. The taxes are collected on his income, net or gross is called as ____________________
15. ________________ is the advance payment of sums due at a later date.
16. ______________denotes the sum paid for a specified quantity and quality for any goods.
17. The average and marginal products are derived from __________________

18. Inputs which get consumed or transformed into products ae known as ___________________

19. The term "agribusiness" was used for the first time by ____________________of Harvard University in 1955.

20. The Line which defines all the possible combinations of two commodities which would yield an equal revenue or income _______________.

21. If the services of some resources (e.g. labour, building, etc.) are not used, those cannot be stocked and such services are known as ___________________

22. The determination of least cost combination of resources is concerned with ___________________type of production relationship.

23. Most of the small businesses can be classified as the following types (Mention):

    1. _________________ 2. _________________

    3. _________________ 4. _________________

    5. _________________ 6. ___________________

    7. _______________.

24. ____________________type of production relationship is explained by the principle of factor substitution or principle of substitution between inputs.

25. The demand for agricultural products (*e.g. food*) is __________. (Inelastic/constant/Elastic)

26. The goal of Product-Product relationship is ________________.

27. The turnover rate in agriculture or forestry is relatively _________ because the production process requires a period of several years. (slow/constant/fast)

28. The resources which are not used in one period of production can be stored for a later period are known as ______________ resources.

29. ___________________type of production relationship is explained by the principle of product substitution and law of equimarginal returns.

30. Rate of capital turnover is the ratio of ________________ and ________________ expressed in a percentage.

31. Time required for a resource to be completely transformed into a product is referred to as________________________.

32. Mention the types of Product-Product Relationships (Enterprise Relationship)

    1.________________ 2. ____________________

    3.________________ 4. ____________________

    5.________________

33. *"Primary force of production"* is the underlying cause of many of the chief difference/s between agriculture / farm forestry and industry such as i. ________________ ii. ________________

    iii. ________________

34. Factor-Product relationship guides the producer in making the decision ____________ to produce? (What/How much/How)

35. In an enterprise relationship of Product-Product Relationships the production of silk cotton (*Ceiba pentandra*) lint and its seed is an example for ________________relationship.

36. In ________________principle the selling price must cover the average variable cost (AVC) to continue production in the short run.

37. The most common method of computing depreciation for livestock in the early years of life, i.e. in the appreciating phase is ________________.

38. A new tree harvesting machine costs Rs.64,000 and expected to last 15 years and the junk value of that machine after 15 years would be Rs.14,000. Then annual depreciation of a machine would be Rs. ________________.

39. ____________________is derived by dividing the sum of working and current assets by the sum of medium term liabilities and current assets which measures the financial safety of the business over an intermediate period of time.

40. Land use efficiency is grouped under ________________ (type) forest efficiency measures.

41. The Four Common Forms of Business Organizations (legal) are:

a) ________________ b) ________________

c) ________________ d) ________________

42. ______________ Curve represents all possible combinations of two resources physically capable of producing the same quantity of output.

43. The Assets which are most liquid and are consumable in a year are called as ______________________.

44. Antagonism relationship is the opposite of the ______________ relationship

45. According the Principle of Costs, the maximum net revenue is obtained, when ______________ is equal to ______________ .

46. ____________ is an outcome of utilization of services

47. Organizational management decisions are sub divided into ___________& ____________

48. Production function measures the relationship between ________ & ____________

49. ______________ is a single convenient unit in production for which output and returns are calculated.

50. The line joining different combination of two inputs which yield the same level of output is called __________________

51. In the first zone of classical production function, increasing returns prevails due to the fact that the quantity of ___________ resources will be in excess of ______________ resources.

52. The value of elasticity of production (Ep) is ________________ in third stage.

53. ______________________ relationship exists, when increase or decrease in the level of one product does not affect the production level of the other product.

54. ________________________is a form of business in which two or more people operate for the common goal of making profit.

Q. Define the Following

1. Production Possibility Curve (PPC)
2. Isoquants
3. Franchising
4. Stock exchange
5. Farm production economics
6. Transformation (Production period)
7. Depreciation
8. Forest accounts
9. Fixed assets
10. Current assets,
11. Net Capital Ratio
12. Shareholder
13. Working Ratio
14. Production
15. Retailing
16. Credit
17. Forest business
18. Stumpage value
19. Yield taxation
20. Corporation
21. Average cost
22. Forest capital
23. Multiple use forestry
24. Discrete function
25. Forest Management
26. Antagonistic enterprises
27. Farm production economics
28. Depreciation
29. Joint product enterprises

30. Stockholder
31. Current Ratio
32. Turnover
33. Stock market / Equity market
34. Trade
35. Sole proprietorship
36. Partnership
37. Project Management
38. A Cooperative
39. Forest taxation
40. Loan / Debt
41. Accounting Period
42. Marginal cost
43. Fixed assets
44. Ridge line
45. Forest capital
46. Marginal physical product

Q. Differentiate the Following

1. Resources Vs Resource services
2. Fixed resources Vs Variable resources
3. Flow resources Vs Stock resources
4. Fixed costs Vs Variable costs
5. Continuous function and Discontinuous function
6. Short run production function and Long run production function
7. Fixed Costs and Variable costs
8. Intensive forestry Vs Multiple use forestry
9. Fixed assets Vs Current assets
10. Long duration liabilities Vs Current liabilities
11. Working Ratio Vs Current Ratio

12. Farm Business Cash Record Vs. Personal Cash Record
13. Short-term loan Vs Long term loan
14. Complementary enterprises Vs Supplementary enterprises
15. Short run production function and Long run production function
16. Factor-Factor Relationship Vs Product-Product Relationship
17. Working assets Vs Fixed assets

Q. Answer the Following

1. Write the scope of forest business in India.
2. Explain the types of small businesses.
3. List the forms of business organization and write the factors affecting forms of business organization.
4. Write the objectives of production economics.
5. Define the Fixed costs and Variable costs.
6. Define the Resources and Resource services.
7. Explain the Factor-Factor relationship.
8. Define the Stock Market and Stock Exchanges. Write the major stock exchanges in the world.
9. Explain the Product-Product relationship.
10. What is production function? Explain the types of production function.
11. Define the principle of variable proportions or laws of returns.
12. Write the characteristics of a good forest (farm) record book.
13. What are the special problems in Computation of Depreciation?
14. What do you mean by Net worth statement? Explain the classification (groups) of assets and liabilities.
15. Write the uses of Forest (Farm) Financial Accounts.
16. Explain the different types of forest efficiency measures.
17. What do you mean by Net Worth Statement? Define the types of Assets and Liabilities
18. Define the different types of privately owned business.

19. What is Project Management? Write the fundamentals of project planning and management.
20. Define the different types of privately owned businesses.
21. Define A Co-operative Business? Write the different types of Co-operatives.
22. Write the Basic Principles of A Co-operative.
23. Write the Economic Functions of the Stock Market.
24. What are the properties of Iso-quant?
25. Explain graphic method to determine the least cost combination.
26. Write a short note on marginal rate of technical substitution.
27. Different types of laws of returns with figures.
28. Types of production functions with equation and definitions of various terms
29. Factors of production and explain all types.

Q. Explain the Following Questions in Detail

1. Explain the characteristics of forest business management in detail.
2. Write the principles involved in forest business management and Explain.
3. Define the production function. Explain the major types of production relationships.
4. What are the difficulties in forest accounting? Mention the advantages of forest records and accounts.
5. What do you mean by depreciation? Explain in detail the most common methods of computing depreciation.
6. Define the loan. Explain the detailed classification of loan.
7. Describe the Product-Product Relationship and explain the types of Product Relationship.
8. Explain the treatment and classification of special transactions for forest financial accounts. What are the different types of forest efficiency measures?

9. What do you mean by A Co-operative Business? Write the Basic Principles of A Co-operative. Explain the different types of Co-operatives.
10. Justify forestry as an independent business enterprise. Write the classification of business.
11. What is project planning and management? Explain in detail about project cycle and its steps.
12. Explain different types of laws of returns with examples.
13. Explain different methods to find out least cost combination of inputs.
14. Explain different type of product-product relationship.
15. What are the different types of enterprise relationship (with examples) do you studied? Explain the algebraic procedure to find the profitable combination of enterprises with figure.
16. Define Law of Diminishing Marginal Returns (LDMR), indicate different zones of classical production function with a neat diagram and bring out the inter-relation between production concepts.

❑❑❑

Appendices

Appendix 1- Terms used in Forest Business Management

A choice indicator : A choice indicator is a yardstick or an index or a criterion indicating which of two or more alternatives is optimum or will maximize a given end. The most desirable combination of products or factors cannot be determined without a choice indicator, which moves production problem from the field of physical sciences to the field of economics. e.g. Price ratios, substitution ratios etc.

Breeding Livestock Enterprise Output : Breeding Livestock Enterprise Output is the value of the secondary product (eg milk, wool, eggs) plus closing value of same, plus progeny sales, and transfers-out of progeny and closing value of progeny, less herd depreciation. Normally, Enterprise Output relates to the recording year, but where the production cycle is less than or greater than one year, it can relate to that period, if so desired.

Cash Flow : Cash Flow is movement of funds into (and out of the business over a defined period and is represented by Receipts and Payments.

Closing valuation is the value of the same items at the close of the accounting period.

Complete Cash Record : It includes expenses on all items both personal and business and income from all sources recorded according to source types.

Concessional loan : A concessional loan, sometimes called a "soft loan," is granted on terms substantially more generous than market loans either through below-market interest rates, by grace periods or a combination of both. Such loans may be made by foreign governments to poor countries or may be offered to employees of lending institutions as an employee benefit.

Consumption Loan : Any loan advanced for purpose of other than production is known as consumption loan. e. g. Medical expenses,

Continuous function : Continuous function can be explained by response of yield to fertilizer or seeds where the doses or levels of inputs and output can be split up into small units. e.g. Fertilizers can be applied to a hectare of land in quantities ranging from a fraction of a kilogram upto hundreds of kilograms. Labour is also quite divisible because it can be used in minutes or in hours.

Contract rearing : An arrangement whereby the owner of livestock pays a fee to a separate farmer to rear or keep the livestock.

Corporation : An artificial being created under state law. A corporation is a separate business entity distinct from its owners, who are called shareholders because they own shares or interests in the corporation. The major characteristic of the corporate form of business organization is this sharp line of distinction between the business and the owners. The corporation is a separate legal entity as well as a separate tax-payer.

Cost of sales : Cost of sales is opening valuation plus purchases minus closing valuation.

Crop Enterprise Output : Crop Enterprise Output is the total value of the crop produced; it equals the value of sales from the crop plus the market value of any part of the crop used on farm, including straw.

Current assets : They are the most liquid assets and are consumable in a year. Examples: seeds, fertilizers, cash on hand, bills receivable, livestock for sale, etc.

Current liabilities : Repayment of these liabilities may be demanded at once.

Current Ratio : It is obtained by dividing the current assets by the current liabilities. This ratio measures the degree of immediate solvency of the business.

Demand loans : Demand loans are short term loans that are atypical in that they do not have fixed dates for repayment and carry a floating interest rate which varies according to the prime lending rate. They can be "called" for repayment by the lending institution at any time. Demand loans may be insecured or secured.

Direct costs : Direct costs are those costs, variable and fixed, that are readily allocated to an enterprise. Indirect costs are those costs that are not readily allocated to an enterprise. Examples are costs incurred for overhead machinery, e.g. hedge trimmers, depreciation, repairs to general purpose buildings, and general expenses, such as stationery.

Discontinuous or discrete function : Discontinuous or discrete function is obtained for input factors or work units which are used or done in whole members, such as one ploughing or a number of ploughings. One can only shift from one point to another. In discrete functions the alternative decision points get limited, otherwise same method of analysis is followed in both continuous and discrete production functions.

Distribution businesses : This classification refers to those businesses, which do not make anything but which bring the goods and services to the consumer or user. This includes such activities as packaging, labelling, transporting, refrigerating, freezing, processing, storing, and performing any service necessary to prepare the goods or to provide the service to eventual consumer.

Double entry system : It is a method of recording each transaction in the books of accounts in its two fold aspects, i.e. two entries are made for each transaction in the same set of books, one being a debit entry and the other a credit entry.

Expenditure or expenses : Expenditure or expenses is Payments adjusted for creditors at the beginning and end of the accounting period (i.e. with opening creditors deducted from payments and closing creditors added).

Farm Accountancy : Farm Accountancy is defined as the art as well as the science of recording in books business transactions in a regular and systematic manner so that their nature, extent and financial effects can be readily ascertained at any time of the year.

Farm Book keeping : Farm Book keeping is known as a system of records written to furnish a history of the business transactions, with the special reference to its financial side.

Farm Business Cash Record : It is a record of all cash receipts from and all cash expenses incurred on the farm business. The farm business cash record omits all personal cash expenses and all non-farm cash income. The objective of this record is to measure changes in the farm business only as far as cash transactions are concerned. The business cash record can be maintained in a simple from or a classified form.

Farm Business Income : Farm Business Income (sometimes referred to as Farm Business Profit) is Total Farm Gross Margin less the sum of the Fixed Costs incurred, before any charges for unpaid labour or notional rent on owner occupied land. Alternatively, it is Total Farm Output less the sum of Variable and Fixed Costs.

Farm Diary : Diary is meant for daily recording of all important events on the farm. It may include weather reports, any unusual events of the day, as well as cash receipts and expenses from all resources.

Farm Executive : Farm Executive carries out the orders of the management.

Farm Manager : Farm Manager is a person who manages or supervises the business according to the instructions of the entrepreneur. He is hired to manage the business and he is not generally responsible for any gain or loss to the business.

Financial Accounting : These records and accounts relate to the operations and activities of the farm business as a whole. Records and accounts under this method set forth income, expenditure, and profit or loss of the farm business as a whole. They reveal the aggregate effect of all the farm undertakings, such as crops growing, dairy, poultry, etc. It gives an overall picture of the entire business and is simpler than cost accounting.

Fixed assets : Fixed assets are of the nature that it is difficult to convert them into cash to meet any current obligations. Examples: land, buildings, etc.

Fixed Costs : Fixed Costs are those costs which either cannot readily be allocated to a specific enterprise or do not vary with small changes in the scale of the individual enterprise. Examples of Fixed Costs are labour (including payments in kind), machinery repairs and depreciation, rent and rates, general expenses, interest.

Fixed Resources : Level of some resources such as buildings, machinery, implements is fixed over a planning period irrespective of the level of enterprise(s) taken up. These are known as fixed farm resources.

Flow Resources : There are some resources, if their services are not used, these cannot be stocked. Services are forthcoming like a flow. These services are labour, buildings, etc. If the services of a labourer are not used today, they cannot be stored until next day, month or the year. Such services are known as flow services.

Full cost accounting : Under this, accounts show income and expenditure and profit or loss of the farm business as a whole as well as of each department, each farm product or commodity as a separate unit. An account is kept for each enterprise or department and all income and expenditure pertaining to the individual farm enterprise is credited or debited to its respective account. This method is useful especially for large farm holders, who can afford the necessary labour or money for keeping such accounts.

Full time equivalent : Full time equivalent is the annual hours of a full time worker (i.e. 1 FTE is 1900 hours per annum).

Grants and subsidies : Grants and subsidies (revenue, not capital). This includes payments from agri-environment schemes. It also includes the single payment, unless recorded elsewhere in the account. Capital grants and subsidies (in respect of new buildings or machinery for example) are excluded from here and recorded later in the account, usually within depreciation.

Gross profit : Gross profit is sales plus grants and subsidies plus sundry revenue plus single payment, less cost of sales.

Inputs : Inputs are resources used in the production process, eg feed, materials, labour and machinery, measured in physical or financial terms.

Interest : Interest comprises financing charges (excluding repayment of capital) paid on loans, overdrafts, hire purchase and finance loans used by the farm business. It is net of any interest earned when these accounts are in surplus.

Intermediate liabilities : These liabilities can be deferred or postponed for the present but fall due within the year. Examples are promissory notes and medium term loans.

Isoquants : An isoquant is a convenient method for compressing three dimensional picture of production into two dimensions. An isoquant represents all possible combinations of two resources (X1 and X2) physically capable of producing the same quantity of output. Isoquants are also known as isoproduct curves or equal product curves or product indifference curves.

Joint Product enterprises : Joint products are produced through a single production process and one of the products cannot be produced alone but must be accompanied by one or more of other products. e.g. Cotton lint and cotton seed, Paddy and straw, etc.

Limited Partnership : A special form of partnership permitted by state law to have one or more partners whose liability for partnership debts and obligations is limited to their investment in the business. A limited partner is just an investor. If a limited partner participates in management, then liability will exist for all partnership obligations like a general partner. A limited partnership must have at least one general partner who handles the management of the business and who is fully liable for all partnership debts and obligations.

Loan : Loan is the certain amount provided for certain purpose on certain condition with some interest and which should be repaired within a stipulated period. Loan is nothing but money borrowed that is usually repaid with interest.

Long duration liabilities : These liabilities are those which do not require repayment during the current the accounting period. Example is long-term loan.

Long run production function : Those input-output relations which permit variation in the input of all factors (none is fixed) can be termed as long run production function.

Long-Term Loan : In this credit period required is more than 5 years up to 15 to 20 years. This loan is meant for leveling of land designing of well, purchase of machineries, etc.

Management : "**Management**" means the act or art of managing.

Managerial Input : Managerial Input is activities that are involved in making decisions in farm business policy. Examples are benchmarking, preparing or interpreting budgets and cashflows, choosing farm enterprises, choosing non-agricultural activities, choosing husbandry techniques. 'Routine' administrative tasks, such as completing support payment claim forms, completing census or survey forms, ordering inputs and paying bills are not included.

Medium-Term Loan : These loans are extended for a period 15 months to 5 years. These loan usually repaid in installment. It is needed for purchase of implement, electric motors, cattle, bunds and other land improvement repair of well, etc.

Net Capital Ratio : It is a ratio between total assets and total liabilities and it is worked out to measure the degree of financial safety over a period of time by comparing the present position of the business with that on some previous date.

Net Cashflow : Net Cashflow is the difference between Receipts and Payments.

Net field rent : Net field rent is the actual rent paid on tenanted land plus imputed rent on tenant's improvements plus imputed rental value (notional rent) of owner-occupied land.

Net Margin : Net Margin of an enterprise equals gross margin less adjusted fixed costs allocated to the enterprise.

Net profit : Net profit is gross profit less overheads less deprecation plus profit (or loss) on sale of assets.

Opening valuation : Opening valuation is the value at the start of the accounting period, of livestock, crops and produce in store, bought and home-grown feeding stuffs and fodder, seeds, fertilisers, pesticides and tillages (that is cultivations and newly sown or growing crops). Livestock, crops and produce are valued at the cost of production or market value after taking account of marketing costs.

Overheads : Overheads (or fixed costs) are expenditure or expenses that cannot readily be traced to processes that produce particular single products. It includes the following: wages and salaries (including casual labour, although note that in management accounts casual labour is often treated as a variable cost), machinery repairs, fuel and oil, contract charges, other machinery expenses (includes renting or hiring machinery, vehicle tax), rent, rates (including drainage charges), property repairs, power, electricity and heat, professional fees, interest payments and bank charges, insurance (not including insurance for labour) water charges, subscriptions, advertising, telephone charges, pest clearance, travel and subsistence (farm share), stationery.

Ownership Charges : Ownership Charges are those General Overheads appropriate to the owned land, buildings and fixed equipment.

Paid Managerial input : Paid Managerial input is the value of managerial input undertaken by a paid manager or worker.

Partially Liquidating Loan : The income generated through these burrower will help to plant of the loan component only. e. g. Dairy loan, Medicinal plant cultivation loan, Tractor loan, Sawmill loan, forest based cottage industry loan, etc.

Personal Cash Record : It is strictly a record of all personal expenses of the farm family, excluding all the farm costs and usually a recording of the cash income of the family from all sources, regardless of whether the income is from farm or nonfarm sources.

Personal Net Cashflow : Personal Net Cashflow is the net flow of funds relating to the nonfarm items of Receipts and Payments. These include funds withdrawn for personal use and tax payments, and also income received from non-farm sources.

Planning horizon : Planning horizon is defined as the period covered by a particular plan or a firm's planning cycle. In general, its length is dictated by the degree of uncertainty in the external environment: higher the uncertainty, shorter the planning horizon.

Product: Products are the result of the use of resources or services of resources: examples are wheat, maize, milk, eggs, etc.

Production businesses: This classification includes all types of production including agricultural production of crops and livestock, as well as forestry.

Production is a process of transformation of certain resources or inputs like land, labour (human), and bullock, seeds, fertilizer, irrigation water into products like wheat, paddy, milk.

Productive Loan: This loan is supply to increase the productivity of crop. These are also known as short-term loans/crop loan. These loans are reply able within period ranging from 6 to 18 months in lump sum.

Professional services: Some type of services, in order to protect the public, requires considerable training on the part of those offering the service. Usually that professional services must have a formal education and rigid examinations before receiving licenses to offer their services to the public. Examples of those offering services are investment brokers, insurance agents, etc.

Profit (or loss) on sale of assets: Profit (or loss) on sale of assets is the profit or loss on the sale of machinery, glasshouses,

Purchases: Purchases are the value of goods and services bought, for cash or on credit. The value excludes VAT if VAT is re-claimed. Purchases covers purchased livestock, bought feed, fertiliser, chemicals, vet and medicine costs and sundry items, such as baler twine, silage wrap, ear tags, and packing materials.

Rearing Livestock Enterprise Output: Rearing Livestock Enterprise Output is the value of finished sales plus value of store or unfinished sales plus value of transfers-out plus closing valuation less purchases less value of transfers-in less opening valuation.

Receipts: Receipts are monies received during the accounting period for the sale of goods1, the provision of services, and from revenue grants and subsidies (but not capital grants and subsidies).

Revenue (or turnover): Revenue (or turnover) is Receipts adjusted for debtors at the beginning and end of the accounting period. (i.e. with opening debtors deducted from receipts and closing debtors added).

Revenue (or turnover): Revenue (or turnover) represents the total trading Receipts for the accounting period, adjusted for debtors at the beginning an end of the accounting period (i.e. with opening debtors deducted from receipts and closing debtors added).

Self Liquidity Loan : The income generated through the loan help the farmer to repay the entire amount in the same year of obtaining loan.

Share farming : Share farming is effectively one business run by two people. Each party provides capital (e.g. livestock, machinery) or resources (e.g. labour) into the arrangement. Unlike contract farming, there is no transfer of money between the two parties. Profit from the sale of produce is divided between the parties at a pre-arranged rate.

Shareholder or **stockholder** : A shareholder or stockholder is an individual or institution (including a corporation) that legally owns a share of stock in a public or private corporation. Shareholders are the owners of a limited company. They buy shares which represent part ownership of a company.

Short and long run production function : Production function which relates to factors and products, where some resources are fixed can be termed as short run production function.

Short-Term Loan : These loans are paid back within period ranging from 5 to 18 months. All crop loans are said to be short term loan, these loans are usually repaid fully at one time in a current production year. This loan can be used for purchasing seeds, land protection measures, taxes, rent and other current farm expenses.

Single Enterprise Accounting : Accounts can be kept for important selected enterprises or a particular aspect of an enterprise, such as production of milk, sugarcane, wool, eggs, etc. This is only a restricted from of cost accounting. It is a simpler method and takes much less time but is not so accurate and complete.

Single Entry System : This is the system which ignores the double effect of transactions. Only personal accounts of debtors and creditors are kept and impersonal accounts are ignored altogether. In fact, any method of accounting which falls short of double entry, may be called as single entry system. Therefore, it is relatively imperfect and its results are less reliable and its accuracy cannot be tested by means of a trial balance which is possible under the double entry system alone.

Sole Proprietorship: A one-person operation. The business may have a number of employees or hired persons; but the proprietor owns, runs, and manages the business.

Standard person-day : Standard person-day represents 8 hours work by an adult worker under average conditions.

Stock Resources : Some resources such as seeds, fertilizer, feeds are entirely used up in the production process. If these resources are not used in one period of production, they can be stored for a later period. These resources are known as stock resources.

Stocking Density : Stocking Density is Grazing Livestock Units per forage hectare. This may be calculated separately for each type of grazing livestock enterprise (where the necessary allocation of Forage Area is both possible and useful). Stocking density may also be measured in terms of Livestock Unit grazing days.

Subsidized loan : A subsidized loan is a loan on which the interest is reduced by an explicit or hidden subsidy. In the context of college loans in the United States, it refers to a loan on which no interest is accrued while a student remains enrolled in education.

Sundry revenue : Sundry revenue is any general revenue, including income from rent, way leaves, hiring out machinery or labour, income from non agricultural diversified activities, such as recreation, tourist accommodation, catering, interest received. Insurance receipts are also included, if they relate to loss of income (eg crop damage) rather than loss of capital assets (such as damage to machinery).

Tax-Option Corporation : A creation of federal tax law. A corporation in all aspects except that the corporate entity pays no income tax because each shareholder owner reports his or her share of corporate income for income tax purposes on their individual income tax returns. Tax-option corporations are subchapter corporations.

Transformation or Production Period : Time required for a resource to be completely transformed into a product is referred to as transformation period or the production period. The production period varies with resources.

Working assets : These are more liquid than fixed assets, such as farm machinery, equipment, etc.,

Working Ratio : It is derived by dividing the sum of working and current assets by the sum of medium-term liabilities and current assets. It measures the financial safety of the business over an intermediate period of time.

❑❑❑